# DRUGS UNLIMITED

*Mike Power*

# Drugs Unlimited

### The Web Revolution That's Changing
### How the World Gets High

## WITHDRAWN

Thomas Dunne Books
St. Martin's Press
New York

THOMAS DUNNE BOOKS.
An imprint of St. Martin's Press.

DRUGS UNLIMITED. Copyright © 2013 by Mike Power. All rights reserved.
Printed in the United States of America. For information, address
St. Martin's Press, 175 Fifth Avenue, New York, N.Y. 10010.

www.thomasdunnebooks.com
www.stmartins.com

Library of Congress Cataloging-in-Publication Data

Power, Mike (Journalist)
  Drugs unlimited : the web revolution that's changing how the world
gets high / Mike Power. — First U.S. Edition.
      p. cm.
  Includes bibliographical references and index.
  ISBN 978-1-250-05471-5 (hardcover)
  ISBN 978-1-4668-5774-2 (e-book)
  1. Drug traffic.   2. Internet marketing.   3. Electronic commerce.
4. Computer crimes.   I. Title.
  HV8079.N3P69 2014
  363.450285'4678—dc23

                                                        2014022130

St. Martin's Press books may be purchased for educational, business,
or promotional use. For information on bulk purchases, please contact
Macmillan Corporate and Premium Sales Department at 1-800-221-
7945, extension 5442, or write specialmarkets@macmillan.com.

First published in Great Britain under the title *Drugs 2.0* by Portobello Books

First U.S. Edition: October 2014

10   9   8   7   6   5   4   3   2   1

To Sasha Carolyn Dunn

# Contents

# A Note on Sources

Throughout this book I have made all efforts to anonymize sources, especially for those who have requested it. Where I have quoted material from online bulletin boards such as Erowid, I have used the posters' original names and citations as they appear online.

# List of Acronyms

Advanced Research Projects Agency Network (ARPANET)
Advisory Council on Misuse of Drugs (ACMD)
Chemical Abstracts Service (CAS)
Defence of the Realm Act (DORA)
Drug Enforcement Administration (DEA)
European Monitoring Centre for Drugs and Drug Addiction (EMCDDA)
Forensic Science Service (FSS)
Home Affairs Select Committee (HASC)
Independent Scientific Committee on Drugs (ISCD)
London Toxicology Group (LTG)
National Security Agency (NSA)
Research Chemical Mailing List (RCML)
Serious Organised Crime Agency (SOCA)
Temporary Class Drug Order (TCDO)
United Nations Office on Drugs and Crime (UNODC)

# Foreword

The only constant feature of the internet drugs scene is change – change so rapid that it is hard to comprehend – and impossible to control.

Since this book was published in the UK in May 2013, the symbiotic processes of chemistry, technology and law have continued exerting evolutionary pressures on each other. Many – if not most – of the drugs mentioned in this edition are now illegal, or subject to temporary class drug orders, in the UK, while every government from the US and the EU continues to play catch-up with chemists, users and hackers.

But as recorded – and predicted – in the first edition of *Drugs Unlimited,* every time a drug has been added to the absurdly lengthening list of chemicals governments say citizens may not ingest, a new legal replacement has come on sale just a few days later. Far from convincing people to stop taking drugs, the state's response has succeeded in pushing the market further down the road of adaptation and experimentation.

In its latest World Drug Report, the United Nations acknowledged this extraordinary expansion: 'While new harmful substances have been emerging with unfailing regularity on the drug scene,' it said, 'the international drug control system is floundering, for the first time, under the speed and creativity of the phenomenon.'

There are now more legal drugs on sale than were even dreamed of when the first global drug laws were written. The 1961 and 1971 UN conventions proscribe just 234, whereas in the last four years alone, a total of 243 new legal compounds have been identified. Fifty-six of these appeared between January 2013 and October 2013, at the time this new edition was going into production.

Another significant development in this period concerns the Silk Road, the Dark Web bazaar where dealers and users all over the world encrypted their communications to buy and sell drugs through national and international postal systems using Bitcoin, the anonymous digital currency.

Chapter 10 from the first edition of *Drugs Unlimited*, entitled 'Your Crack's in the Post', was the world's first book-length excursion down the Silk Road, but in October 2013, the site was closed down and its alleged owner arrested. Sure enough, just a few weeks later, a new, more impregnable site reopened, along with around a dozen new anonymous drug markets.

In this new edition, an update of that chapter, including an exclusive interview with the new owner of the Silk Road, follows the original text – but it's a certain bet that events will outrun print schedules and the story will continue to unfold.

We now all accept that the internet has changed our daily lives and our social and psychic landscapes in ways that make the pre-web past seem dreamlike and remote. We also need to acknowledge that the digital age has changed worldwide drug habits just as profoundly, and that the old legal model of drug control is now both unworkable and counterproductive.

Rather than further extensions of prohibition, a more rational response to drug use would be to reduce the harms that drug users face. We now need to use the web – and its founding principles of collaboration and communication – to create new ideas and strategies that will end the unwinnable war on drugs.

Mike Power
London
February 2014

# Prologue

## *Contemporary Chemical Culture*

In 2011 forty-nine brand new psychoactive drugs were invented and advertised for sale on the internet, from which they were bought perfectly legally by curious consumers. In 2012, fifty-seven new drugs could be found for sale online. Their dosages were not always clearly specified, and were sometimes far tinier than the recreational drugs people were used to taking; their effects were undocumented – and yet no law could prevent their sale. Just over a decade before, no new psychoactive drugs had been available in this way. These were not the drugs – like heroin, cocaine and Ecstasy – that people had used for years. Their names were baffling lists of numbers and letters, such as 6-APB, 5-MeO-DMT, 3-MeO-PCP, and made them sound more like laboratory supplies than recreational drugs. But they were being taken by ordinary people, and global discussions about their effects were going on as blatantly as their sale.

Like nearly every other industry, the drugs market has been revolutionized by the web. For a growing number of people it is now the first place they look when trying to source recreational drugs or information about them – especially when faced with the rapid and baffling proliferation of new compounds. Chemists, consumers and criminals all use the internet to share vast amounts of information and exploit globalized manufacturing possibilities. Completely untested compounds are part of an international online market that has become too fast and too complex for any government to control properly. Welcome to Drugs 2.0 – an anarchic free-market world in which drug legislation is being outpaced by chemistry, technology and ingenuity. How did we get here?

This situation was accelerated in some small way by a story I broke in British magazines and newspapers in 2009. My pieces were the first in the world to document the emergence and popularization of mephedrone, an Ecstasy and cocaine substitute that had escaped from an underground online drugs scene. Mephedrone became the first new drug since Ecstasy to hit the front pages of newspapers worldwide, and to prompt questions in Parliament and the American House of Representatives, and the governments of many other countries.

Illegal drugs, including LSD in the 1960s, heroin in the 1980s and Ecstasy in the 1980s and 1990s, have long had a uniquely perturbing influence on the public realm, with the dangers and pleasures inherent in their consumption splitting users and law makers into opposing camps. Mephedrone, though, was completely legal. The new drug was, according to toxicologists, 'two chemical tweaks away from Ecstasy'. Those tweaks were deliberate, and were made to evade drug laws. Mephedrone upended all prior hierarchies and caused huge confusion among many users who considered, wrongly, that since it was legal, it was harmless.

Widely available and hugely popular, mephedrone was the first mass-market 'downloadable' drug, in the sense that it was, uniquely for the mass market, originally only available online. It was like a narcotic viral video, a digital diversion to be shared with the click of a mouse. In every sense, it was a radically new game-changer. Mephedrone was the fulcrum, the tipping point that took a clandestine internet drug scene and dropped it, gurning and wide-eyed, right into the high streets of the UK – and then into the wider world. The swift and protocol-busting ban on the new drug in the UK did nothing to eliminate it here or in the EU or the US; it simply handed the market to grateful gangsters who added the drug to their repertoire, and prompted greater innovation in the chemical underground.

Following the intense media attention that mephedrone attracted, and the ensuing moral panic, the new chemical craze of so-called 'legal highs' gained full-spectrum media dominance in a matter of weeks. Newspapers and legislators were shocked, but the situation was as predictable as it was inevitable – if you knew where to look and what you were looking for, and if you'd been looking for long enough.

The online 'research chemical' scene is at the root of this story, and it is from here that mephedrone sprang. Whereas Ecstasy, cocaine, amphetamines, marijuana and tranquillizers have been used by countless people for decades or centuries, or, in the case of cannabis, millennia, and have often been the subject of costly animal and medical trials costing millions of dollars, research chemicals have little to no history of human usage. These new drugs are generally active in minute doses of single or double-digit milligrams (by comparison, a dose of Ecstasy weighs an eighth of a gram, or 125 mg).

Until around 2007 they were used by a few thousand self-defined 'psychonauts', or explorers of inner space, who researched the compounds' effects by browsing scientific literature, or, in some cases, simply by looking at diagrams of molecules, and shared their experiences on online discussion forums. Today, there are probably hundreds of thousands of users of research chemicals, though mephedrone showed how, when the social, cultural and technical conditions are right, new and completely untested drugs can spill from the underground right into the mainstream, winning millions of enthusiastic if uninformed users.

On Sunday 17 June 2012, a poster named Clapham Boy wrote the following, titled '6-apb powder and some MXE', on the Urban75 drugs forum:

What I find so amazing is that you can now go online, order this sort of stuff totally legally (I got 10g of MXE just before the ban on importing & selling it came in, possession remains legal at the moment, and [I] invested in 1g of 6-APB this week after reading a post [here]) and have it delivered recorded delivery the next day. It just makes it so easy compared with tracking down a decent dealer that's not too bloody dodgy, like we had to do when I was younger. It's funny, but whilst I had heard of 'legal highs' a few years back, I just assumed they would basically be crap, and didn't look into it further. It was only after reading an article in the 'i' newspaper, earlier this year, about MXE and its effects that I thought, 'That sounds fun', and decided to google it . . . The whole situation just seems totally mad, and certainly blows a massive hole in the drug laws . . . I expect 6-APB will be banned soon, so I'll have to stock up a bit on that before it happens, and sit back and wait [and] see what new legal highs are produced to replace those banned.[1]

Clapham Boy's casual, almost throwaway attitude to the consumption of new and powerful drugs is commonplace now, yet it is a perspective that is seldom encountered by policymakers or police. Conventional academic research and government-sponsored investigations into attitudes and use patterns are being supplanted in their authority by the unmediated voices of users themselves, as social networks become central to the daily experience of a new generation of drug users.

The chemicals Clapham Boy mentions, MXE and 6-APB, have existed on the street for less than a couple of years, and both were given life on the web. MXE is methoxetamine, a drug closely related to the banned anaesthetic ketamine: a few additions to ketamine's molecular structure have rendered the resulting compound legal in most countries. Both drugs can send users into bizarre internal spaces, imaginary realms where mind and body are dissociated from each other, and where the only limits to the experience are those of the imagination. Methoxetamine is perhaps not physically addictive, but it can be psychologically so seductive that some users have posted in web forums that they have dosed on it for days, with disastrous consequences. It is active at very small doses of around 10 mg. Each gram provides up to 100 doses and, until it was banned in 2012, cost as little as twelve pounds – meaning each hit cost around twelve pence. From 2010 to 2012, methoxetamine was sold entirely legally on hundreds of websites.

The other drug mentioned, 6-APB, is similar to MDMA, or Ecstasy. It costs around twenty-five pounds a gram, and its effects are broadly like those of its parent chemical – it's mildly hallucinogenic, euphoric and emotionally psychedelic; users feel more friendly and open, and their enjoyment of music is enhanced. Each gram contains ten doses, and costs around thirty-five pounds. It is still, at the time of writing, legal, and in common with methoxetamine has almost no history of human use.

6-APB and methoxetamine are just two of hundreds of new, potent, legal drugs that have become available since mephedrone was banned in 2010. Nothing more than a net connection and a credit card are required for their purchase. The chemists' ingenuity is simply outpacing lawmakers' ability to respond; stimulants, sedatives, psychedelics, cannabis substitutes, drugs with similar effects to heroin – all chemical life is here.

Where did these and the hundreds of other new drugs come from? How were they conceived of and manufactured? How can substances so completely untested and unregulated escape legal prohibition? Why are people willing to take them? And what happens when the human desire for altered states is transposed onto the virtual shopping space of the net? Four years ago, I set out to answer these questions.

My own interest in drugs stemmed from many enjoyable experiences that I had had more than twenty years previously, when Ecstasy first appeared in the UK. My responses to the drug and its surrounding culture, and those of everyone I knew, were markedly different from the media's representation of them. Where tabloid editors saw a moral threat, we saw a positive moral choice: rejecting the aggressive and misogynistic culture of tacky nightclubs and renouncing a politics that told us society did not exist, we danced together in fields and warehouses to music whose lyrics often urged unity and peace. That utopian delusion was arguably no more extreme and certainly far less objectionable than its cultural counterpoint, which suggested that violence and drunken boorishness were not only acceptable, but preferable.

In the intervening decades, I discovered, drug culture had become far more complex and widely distributed, and much more dangerous. Media representations of drug use and users were, however, still as shrill and inaccurate as they had been in my youth, and I felt qualified to investigate the culture and its history and report back on it, to document the truth of the matter.

To investigate this world, this underground scene within the already subcultural world of drug use, I registered on dozens of web forums dedicated to drugs and discovered what new substances were available, and where they came from, chemically, historically and geographically. I found sources for the drugs, and verified them. I learned about their effects through interviews and lengthy discussions with users and dealers. I lurked on underground web forums on encrypted anonymized internet connections, posing as an international buyer for bulk quantities of drugs both legal and illegal. Immersing myself in this way has enabled me to gather information that is generally missing in traditional reporting around drugs – and *always* missing when writing about the new drugs scene.

I also met with and interviewed dozens of scientists,

toxicologists, doctors, nurses, chemists and police, to hear how the phenomenon has affected their lives and industries.

While I recorded this emerging social trend over the last four years, it became increasingly apparent that I was witnessing and documenting the initial stages of a major shift in many international drug users' habits. In brief, the sale and use of new drugs are growing, and there is no evidence to suggest that this will stop. This book plots the changes in the chemical culture, and makes plain their causes and consequences. It traces the story back to humankind's earliest use of drugs, outlines the history of their legal prohibition, looks at the dramatic emergence of psychedelics, when drugs changed popular culture for the first time, and explains how, in the modern era, the internet has changed the story profoundly. I examine how we have reached this extraordinarily novel situation – of powerful new drugs being sold legally online – and reveal how new uses of internet technologies are making it possible to buy even banned drugs such as heroin, crack cocaine, crystal methamphetamine, LSD and Ecstasy from secret websites with little fear of detection. In short, I ask and answer the question: Have drugs won the drugs war?

Since 2009, when mephedrone gained popularity, many parts of the drugs market have atomized, and with this increase in variety and novelty come new dangers, and an urgent need for action. There is no safe path forwards – but there's now simply no way back to the way things were. The genie is out of the test tube.

But first, let's go back, right back to the earliest documented occasions of humans deliberately changing their state of consciousness. Because while culture and social conditions might change, neuropharmacology and the human desire for transcendence, stimulation – or oblivion – are essentially immutable.

# 1

## *Vegetable to Chemical*

She was somewhere around Pang Mapha, thirty-one miles from the Salween River, when the drugs began to take hold. In Spirit Cave in north-western Thailand in about 9000 BC, an early human from what is today Vietnam was chewing on some areca nuts, betel leaf and slaked lime, holding the bitter mix in her cheek and absorbing the stimulating alkaloids through her gums. She later spat out the residue, and that spent wad of vegetable matter was found in the mid-1960s, by American archaeologist Chester F. Gorman on an expedition to the area. It is the world's first known use of psychoactive drugs, discovered fully 11,000 years after the event.[1]

To reach the first documented psychedelic experience, and indeed the first example of someone sharing written information about drug effects, we must jump forward over seven thousand years. In around 2700 BC, Chinese Emperor Shen-Nung described his experience of taking cannabis: 'Medical cannabis. Stop eating. Let go. Eat more. You will see white ghosts walking around. And eat long enough, you will know how to talk to the Gods.'[2] Emperor Shen-Nung, or 'Divine Farmer', was a mythical hero in Chinese culture, revered even today by practitioners of traditional medicine. He is the father of Chinese pharmacology and is credited with teaching his subjects how to grow food. His pharmacopeia, quoted above, also mentioned herbal cannabis as a curative, a citation widely accepted by scholars as being the first reference to the plant's medicinal qualities.

The adventurous emperor did not stop at cannabis. Shen-Nung ingested hundreds of other plants to establish their toxicity – or, from another perspective, their powers – and he is said once to have consumed seventy toxins in a single day. But what he did with his mind and his body was of no concern to the authorities. The innate desire of humans to ingest substances that allow us to explore our minds, or experience pleasure, or search for new knowledge and experiences wasn't always controlled by law.

The use of plants and natural materials that affect our consciousness is documented historically in most corners of the globe. Tea, coffee and coca, tobacco and betel, guarana and khat are all natural stimulants that affect the central nervous system, and all have been used for millennia. Dozens of mushroom species, marijuana strains, cacti, seeds and barks, the latex produced by certain flower pods – whether psychedelics or sedatives or deliriants – have been used to induce altered states. From the dawn of human history, plant specialists and medicine women and men with expert knowledge of their effects have been revered figures, especially in pre-industrial societies. Their work crossed the boundaries between the modern disciplines of psychiatry, general practice, religion and magic.

Today the use of substances that change consciousness is proscribed by most nations on the earth, but around five per cent of the planet's inhabitants continue to use drugs that are now deemed illegal. The journey that brought us to this curious point in history is surprisingly short: laws controlling the use of drugs are less than 150 years old. Most recreational drugs at one point had a legitimate purpose. The two most popular at the turn of the century, cocaine and opium, were mainly used as medicines, and the first drug laws were written in order to prevent the dangers of death or addiction arising from their misuse. The number of drugs abused in Europe and the US at the turn of the twentieth century was minimal, and all of those were plant-derived. Psychedelics were, at this time, limited to natural products: marijuana, and psilocybin-containing or 'magic' mushrooms, and, overseas, mescaline, as found in peyote and other cacti. The latter two drugs were used mainly in religious or ritual contexts in Central and Latin American agrarian societies; they were not often imported or traded, certainly they were not easily available. That reality was reflected in the laws of the era: most early drug legislation was not designed to prevent recreational use, which did not exist in any meaningful or threatening way.

In 1908, Britain passed its first anti-drugs legislation when the Pharmacy Act of 1868 was amended to regulate provision of opium found in medical products, with the aim of preventing poisonings or suicides. Preparations containing opium were henceforth required to be labelled as poisons, although their sale and consumption were not limited. The first American drug law was also opium-related: the government passed a ban on the smoking of opium in 1875, specifically written to target immigrant Chinese citizens in San Francisco and their supposed moral turpitude. The International Opium Convention, the world's first international drug control treaty, was passed in the Hague in 1912.

In 1916 Britain's Defence of the Realm Act (DORA) placed both opium and cocaine under the control of the Home Office. Both drugs were being used as pain relief medicines and anaesthetics during the First World War, and were scheduled in order to protect supplies for the injured and ill. The DORA also aimed to curb the use of cocaine by soldiers in London on leave from war service.[3]

Britain's Dangerous Drugs Act of 1920 went further and limited the production, import, export, possession, sale or distribution of opium, cocaine or morphine to licensed persons. At this point anti-drug laws were easy enough to write and easier yet to enforce. It is a simple matter, legally and chemically speaking, to outlaw a drug contained in a plant, even if such moves are felt by some to be philosophically hard to justify. But this situation was not to last, because drug-making was soon to become inextricably linked with the laboratory.

Opium, magic mushrooms, mescaline-containing cacti, marijuana and, to a lesser degree, cocaine, are all essentially natural drugs. They are extracted or concentrated from plant sources and, with the exception of cocaine, there is no laboratory work involved in their manufacture. The active ingredient of cocaine is extracted from coca leaves, and the extract is concentrated and then combined with an acid that makes the drug rapidly absorbable by the body, but it undergoes no significant molecular change. The other drugs listed above are simply gathered and eaten or smoked. Drugs such as these grow easily only in certain geographical and climatic conditions, and their processing tends to take place in the countries of production.

Synthetic drugs, by contrast, are made in laboratories, and for every one of them it is possible to produce a variation on the parent structure; this makes it difficult to write all-encompassing laws banning them because a slightly new structure is always possible. Organic chemists, who work with carbon-based compounds, can reproduce nearly any natural compound, including any of the active ingredients in those traditional, plant-based drugs.

In the early nineteenth century, scientists did not believe it was possible to synthetically produce certain chemicals derived from living organisms. That was proven to be untrue by German chemist Friedrich Wöhler in 1828 when he produced urea, a constituent of human urine, in the lab. 'This investigation has yielded an unanticipated result that reaction of cyanic acid with ammonia gives urea, a noteworthy result in as much as it provides an example of the artificial production of an organic, indeed a so-called animal, substance from inorganic substances,' he wrote in *The Annals of Physics*, heralding the birth of organic chemistry.[4]

All synthetic organic chemical structures are now built in the laboratory in the same way that a builder constructs a house. Basic chemical building blocks – elements – are bonded together in the lab by reacting them with other agents in controlled chemical and physical environments, using heat, acidity and a lack or surfeit of air and water, or any of a hundred other conditions and methods, to produce compounds. From the late nineteenth and early twentieth centuries to the present, pharmaceutical chemists have used the same principles to modify existing drugs and medicines in an attempt to produce variants that are more effective, more potent, or have fewer side effects, and also to produce drugs that are unpatented and therefore possible to commercialize and sell at a profit. Chemist Charles Romley Alder Wright first synthesized diacetylmorphine – known today as diamorphine, or heroin – in 1874 in St Mary's Hospital, London, in a search for a new drug to help wean morphine addicts off the drug. He boiled together a reagent called acetic anhydride with morphine for several hours. This reaction added a new group of chemicals, known as a functional group, to the main morphine skeleton.

There are many functional groups in organic chemistry, and each of them has a different effect on the way the body processes and experiences a drug. In the case of morphine, the addition of two structures made up of two carbons, three hydrogens and an oxygen molecule, known as an acetyl group, made the new drug more fat-soluble, and therefore more of the active ingredient was able to pass through the blood–brain barrier. The blood–brain barrier protects the central nervous system from foreign substances that may injure it, and maintains a constant environment for the brain. Large molecules do not pass easily through it, and the entry of highly electrically charged molecules is similarly slowed. Molecules that are not fat-soluble cannot enter the brain at all. Potency is proportional to efficacy and affinity – a measure of how well a drug binds to a given brain receptor, and its ability to effect a response within the brain and body. By adding this functional group, Alder Wright produced a new drug with completely different effects: heroin, which enters the brain more rapidly and which produces a more euphoric effect than its parent molecule, morphine. It is also even more addictive.

Changing the chemical formulae of drugs, even in a seemingly insignificant way, means their effects can be modulated, amplified, extended, decreased or in some way made different, and potency can be increased or decreased by the addition of functional groups. This is a chemical process called 'ring substitution', since different elements are bonded to the parent drug's chemical rings (see pages 30–33). These new drugs created using ring substitution are called analogues: they are essentially legal versions of banned drugs, deliberately invented by chemists who add or take away a few molecules from illegal drugs and then commercialize them.

The process of creating or unearthing a legal version of a banned drug started as soon as the first internationally binding drug laws were passed. A little after the ink dried on the International Opium Convention in 1912, dibenzoylmorphine and acetylpropionylmorphine – legal alternatives to newly controlled morphine and heroin – became available; they could be considered the world's first designer drugs, or controlled substance analogues.

Drug use in the pre-psychedelic age in Europe and America remained limited to a subculture made up of junkies and the underclass, bohemians and aristocrats, with minimal penetration into the broader culture. But in the 1940s and 1950s, new hallucinogens emerged, soon followed by new stimulants; both were to have profound effects on popular culture and move drugs into the mainstream. And the involvement of the laboratory in their production made it far more of a challenge to legislate against their use.

\*

The emergence of psychedelics into western culture was sudden, unexpected and dramatic, and has had long-lasting consequences.

On 16 April 1943, Swiss scientist Albert Hofmann made a curious decision to resynthesize a compound he had made in the Sandoz laboratories in Basel five years earlier. In 1938 the chemist had been producing a series of compounds related to ergot alkaloids, some of which had been successfully used to stem blood loss in mothers giving birth. Ergot is a kind of fungus that can colonize grains such as rye, and its ingestion can lead to convulsions, delirium, madness and gangrene, since it narrows veins and cuts off blood supplies to extremities. In medieval times, attacks of ergotism were seen as divine punishment rather than the simple chemical consequence of eating contaminated bread.

In his work with these alkaloids, Hofmann was attempting to discover a drug known as an analeptic, which would stimulate the respiratory system, and so, as is common in the field, he produced many slightly different variants of the parent drug – in this case, lysergic acid – in a process known as structure–activity–relationship. 'Thus among other compounds, I synthesized the diethylamide of lysergic acid with the intention of obtaining an analeptic. This compound might have been expected to possess analeptic properties because of its structural relationship with the well-known circulatory stimulant nikethamide,'[5] wrote the chemist.

Hofmann's first experience of the drug was not deliberate: he absorbed a microscopic amount accidentally through his fingertips. Feeling unusual, he left the lab and cycled home. The following week he described the experience:

> Last Friday, April 16, 1943, I was forced to stop my work in the laboratory in the middle of the afternoon and to go home, as I was seized by a peculiar restlessness associated with a sensation of mild dizziness. On arriving home, I

lay down and sank into a kind of drunkenness which was not unpleasant and which was characterized by extreme activity of imagination. As I lay in a dazed condition with my eyes closed (I experienced daylight as disagreeably bright) there surged upon me an uninterrupted stream of fantastic images of extraordinary plasticity and vividness and accompanied by an intense, kaleidoscope-like play of colors. This condition gradually passed off after about two hours.

Baffled as to how the drug could have entered his body in any dosage adequate to cause such extraordinary sensations, Hofmann repeated the experiment deliberately a few days later and again recorded the experience for others to read:

> The notes in my laboratory journal read as follows:
> April 19, 1943: Preparation of an 0.5% aqueous solution of d–lysergic acid diethylamide tartrate.
> 4:20 P.M.: 0.5 cc (0.25 mg LSD) ingested orally. The solution is tasteless.
> 4:50 P.M.: no trace of any effect.
> 5:00 P.M.: slight dizziness, unrest, difficulty in concentration, visual disturbances, marked desire to laugh . . .
> At this point the laboratory notes are discontinued: The last words were written only with great difficulty. I asked my laboratory assistant to accompany me home as I believed that I should have a repetition of the disturbance of the previous Friday.

With that (rather large) quarter-milligram dose of a tasteless white powder, the psychedelic era began, as did the era of synthetic, man-made and recreational drug-taking that persists to this day. LSD is so potent – active at just a tenth of a milligram – that it enabled drug-taking on a scale never before seen. A single gram could dose 10,000 people. Cultural considerations aside, it was this potency and the subsequent potential for profit that so animated the drug culture that was to follow.

First, though, Hofmann's creation was used by psychiatrists convinced that LSD could unlock the mysteries of human consciousness, and in particular of mental illnesses, including schizophrenia. Indeed, the first descriptor for psychoactive drugs – psychotomimetic – was derived from the belief that the drugs temporarily induced psychosis. Scientists in the UK and US were at that time free to test these powerful new compounds on psychiatric patients as they wished, with no interference from the law, as they were not scheduled or defined as having any legitimate medical use. Dr Ronald Sandison, who died in 2010, was an early pioneer in Britain in the clinical use of LSD. In 1952, after visiting the Sandoz labs in Switzerland and meeting Hofmann, Sandison moved to Powick Hospital in Worcestershire and began radical, government-funded work on the psychiatric efficacy of the substance. There, 683 psychiatric patients were treated with LSD a total of 13,000 times, for which many of them received compensation from the NHS in 2002.

In the US, the army and the CIA experimented with LSD as a truth serum as part of its despicable, decades-long Project MKultra, dosing victims unwittingly in a bid to control their minds in that Cold War era.

At first psychedelic drugs were only available to an elite of scientists, travellers or the fantastically curious. In the middle of the century, bohemians and intellectuals in the West became interested in these consciousness-expanding substances. Initially they sought plant-based psychedelics. Later, they would look to the laboratory.

Beat Generation authors such as William S. Burroughs and Allen Ginsberg freed – or warped – their minds using LSD, and any other chemical they could find, including peyote, mescaline and vines from tropical jungles containing DMT, one of the world's most extraordinary hallucinogens. Burroughs and Ginsberg's *The Yage Letters* documents Burroughs' quest to consume the mysterious Yage vine, which he has heard can cure heroin addiction, and inspire telepathy (it does neither). Burroughs had read of the brew's power in a paper by Harvard professor Richard Evans Shultes, the father of modern ethnobotany, and essentially had no idea what this drug was, but gamely set off to Latin America to learn more. His odyssey began with a haemorrhoid operation in steamy Panama City in January 1953; he then travelled to a miserable and rainy Bogotá a month later, where he sought out a university professor, Dr Schindler, who was connected with the American agricultural commission. Schindler directed the author to Putumayo, in the south of the country, to seek out a *curandero*, or shaman. His circuitous route took him to Cali, Armenia, Popayan, then to Pasto on the border of Ecuador to ruminate on leprosy, and to Puerto Limón, near Macoa, where he heard he could meet a *brujo*, or witch doctor. He jumped in a canoe to Puerto Assis, and a young man stole his underpants in a jungle tryst. Burroughs was then arrested, and sent back to Bogotá as his tourist card was out of date thanks to a typo. Undefeated, he returned to the jungle once more on this epic quest and it was not until 15 April – and after a bout of malaria – that he was able to write to Ginsberg, 'Back in Bogotá. I have a crate of Yage.'[6]

(If Burroughs were searching for an excursion into the N,N-dimethyltryptamine space today, for this is the vine's active ingredient, he would need only to google the following phrase: 'Buy DMT vine' and he would have it a few days later. And in some alternate quantum dimension, if Burroughs were alive in 2013, and feeling cautious for once in his life, he could avoid the attentions of the law by simply searching for powerful and legal DMT analogues available, legally, online in the USA. He could send an electronic payment, and have the drug posted to him the next day. And if he had wanted to fulfil his evident lifelong death wish, he could, until January 2011, have ordered an even more powerful chemical in the same family, 5-MeO-DMT, at that point legal in the US, smoked a few milligrams from a glass pipe, and experienced the closest state to death and a consciousness of the void available without his heart actually beating its last.)

The year 1955 marked another watershed in psychedelic history. On 23 June, in a small adobe hut in Oaxaca, Mexico, American investment banker R. Gordon Wasson of J.P. Morgan was handed a dozen mushrooms by *curandera* – or female shaman – Eva Mendez. He ate them. Unknown to the banker, Mendez had given him psilocybin, a chemical named O-phosphoryl-4-hydroxy-N,N-dimethyltryptamine or more snappily, 4-PO-DMT. Psilocybin, a naturally occurring substance, is found in hundreds of mushroom species at many latitudes on the planet. The backbone substance, tryptamine, is one of the two main chemical classes of drugs that can cause psychedelic or visionary experiences in humans. Skilled chemists can produce thousands of variations on the tryptamine form.

The effect of drugs related to tryptamine tends to resemble the magic mushroom experience – though this is to paint it with enormously broad and clumsy brushstrokes. Typical effects are open and closed-eye hallucinations with geometric patterns and visions, amusement, bewilderment, auditory distortion and, for an unfortunate few, thought-looping existential crises or blind panic. In an article in the 13 May 1957 edition of *Life* magazine, Wasson said his mushroom expedition was the culmination of a lifelong quest to understand the links between mythology and toadstools. Whatever his intentions, and whatever the root cause of his motivation (conspiracies abound) he did a fine job of persuading millions of people after him to indulge in a natural psychedelic experience. He wrote:

> We were never more wide awake, and the visions came whether our eyes were opened or closed. They emerged from the center of the field of vision, opening up as they

came, now rushing, now slowly, at the pace that our will chose. They were in vivid color, always harmonious. They began with art motifs, angular such as might decorate carpets or textiles or wallpaper or the drawing board of an architect. Then they evolved into palaces with courts, arcades, gardens – resplendent palaces all laid over with semiprecious stones.

Then I saw a mythological beast drawing a regal chariot. Later it was as though the walls of our house had dissolved, and my spirit had flown forth, and I was suspended in mid-air viewing landscapes of mountains, with camel caravans advancing slowly across the slopes, the mountains rising tier above tier to the very heavens.[7]

The psychedelic experience differs markedly to most previous drug states available to westerners up until that time. Opiates and cocaine sedate or cause euphoria and a sense of increased energy and confidence, but true psychedelics cause a simultaneous ontological explosion and implosion. They have an effect upon the mind that cannot be simply expressed, and which, indeed, can only be truly understood if experienced. And taking psychedelics seems to spur people on to describe, document, discuss and communicate their experiences, generally because they are so baffling and novel. The drugs are not addictive, metabolically, but their effects are so profound that many users return for more.

Wasson was so dumbfounded that he went back to repeat the experience just three days later:

I repeated the same experience in the same room with the same *curanderas*, instead of mountains I saw river estuaries, pellucid water flowing through an endless expanse of reeds down to a measureless sea, all by the pastel light of a horizontal sun. This time a human figure appeared, a woman in primitive costume, standing and staring across the water, enigmatic, beautiful, like a sculpture except that she breathed and was wearing woven colored garments. It seemed as though I was viewing a world of which I was not a part and with which I could not hope to establish contact. There I was, poised in space, a disembodied eye, invisible, incorporeal, seeing but not seen.

Many intellectuals, chemists and other inquisitive people were fascinated by the psychedelic drug mescaline in the 1950s. The active chemical was first isolated from the peyote cactus by Arthur Heffter, a German pharmacologist, in 1897. Ernst Spath, an Austrian chemist, synthesized the drug in the lab in 1919. Mescaline, or 3,4,5-trimethoxyphenethylamine, is the drug that startled author Aldous Huxley into writing *The Doors of Perception*, the 1954 psychedelic classic. Its chemical backbone is phenethylamine, the other category of drug that prompts hallucinations and mystical experiences in humans. Phenethylamine is as malleable in the hands of a skilled chemist as the tryptamine skeleton, and can be manipulated in similar countless ways to produce hallucinogenic stimulant drugs.

Mescaline, which has long been used by indigenous peoples for religious and divinatory purposes, exists naturally in hundreds of cacti species all over the Americas, with a few examples found in the Caribbean. The presence of the extra chemicals on the phenethylamine skeleton changes its subjective effects, and makes humans who take it hallucinate and experience extraordinary and unusual effects, such as a sense of wonder at the natural world, or indeed almost anything, much laughter and intimacy with friends, along with a feeling of peace and, often, profound reflection in the final stages of the drug.

Lawmakers worldwide have forbidden most humans from experiencing its effects, although it has been used for millennia, and its effects have been studied over long periods in traditional user groups and found to be largely harmless, even beneficial.

One of the first and most diverting documentations of the use of mescaline in its natural form was actually made in 1898, in *The Contemporary Review*, one of the oldest English publications in the world, founded in 1866. Havelock Ellis was a British physician and social reformer who became renowned for his a radically objective scientific approach to the study of sexuality. He wrote the first medical textbook on homosexuality, and though some of his stances are anachronistic and even objectionable today – he details man-boy relationships impassively, and was later a eugenicist – his writings contained an unflinching look at hidden aspects of modern life as it accelerated into the Victorian age. He moved in circles that later formed the Fabian Society, Britain's oldest political think-tank.

His 1898 essay for the *Contemporary Review*, 'Mescal, a New Artificial Paradise',[8] brought some of that modern and open-minded attitude to the world of psychoactive drugs. The first of his writings on mescaline, it is an evocative description of the experience and predates Aldous Huxley's musings on the matter by almost half a century. Ellis used his own body as the laboratory of discovery, and his work was neither illegal nor hidden. On Good Friday the previous year, at around 2.30 p.m., Ellis had brewed up some peyote buttons in his rooms in the Temple, London, and swallowed the bitter tisane. Soon afterwards, he started hallucinating.

He described how he was gently assailed by benign, bejewelled visions. Unfamiliar and intricate flower petals and gauzy butterfly wings wreathed his mind, folding and morphing before his eyes, and behind his eyes, for it mattered not whether they were opened or closed, he said. Every colour of the spectrum was present, a vast profusion, and he delighted in this glorious aesthetic overdose as his visions transformed once more into porcelain and lace, lattice-style *mouchrabieh* woodcarvings of Cairo, towering Maori-style archetypes of architecture.

As time passed and the drug's effects diminished, he found

sleep elusive and became fixated on his own legs and the shadows on the ceiling. Thirteen hours later, he closed his notebook after documenting the whole day's events. He soon introduced friends to the experience, who reported extraordinary visions and 'a very marked sense of well-being'.

Almost sixty years later, at noon on Friday, 2 December 1955, a distinguished member of the British establishment had a similar experience. A well-dressed pair of men, with cut-glass accents so refined they could contain a small measure of sherry, sat in an austere, post-war British front room. For one of the men, that would remain the case. The other man was about to witness an exquisite aesthetic transformation and a baffling temporal shift that would change his worldview for many decades to come. Between them sat a microphone, and a TV crew for BBC1's *Panorama* documentary series filming the event. It marks the emergence of psychedelic drugs from the laboratory and the clinic into mainstream culture.

Liberal (at the time; he had previously been Labour) MP Christopher Mayhew steepled his fingers nervously and leant back with an awful, uneasy élan as the BBC's cameras rolled. He was preparing to be administered a strong, 400 mg oral dose of pure mescaline hydrochloride. The clock swung to midday. 'I'm feeling perfectly fit at the moment and as sane as I ever am. And I'll take the drug now,' said Mayhew in his stiff starched collars and brilliantine. An hour later, the molecule had entered the MP's brain, and his curtains, previously a mundane blue, now seemed to be a florid, splendid mass of writhing colour. He was most pleased by their appearance. His face, before a tense knot of feigned confidence, was now lit up with a beaming grin he couldn't quite control. The friend who had given him the drug, the fabulously named Humphrey Fortescue Osmond, was a progressive psychiatrist who claimed his place in drug history by inventing the word 'psychedelic', meaning 'mind manifesting', in correspondence with fellow mescaline fan and author Aldous

Huxley. Today, Osmond would be classed as a drug dealer, though he was serving up highs that were, at that time, legal.

They weren't even considered highs, but rather medicines; Osmond successfully treated many alcoholics with LSD in the 1950s until the drug was banned. Mayhew couldn't have been in safer hands, even if Osmond did proceed to oversee a bizarre interview in which the MP, once the drug took hold, became convinced he was travelling through time. 'Now, I'm off again, Humphrey,' he said. 'In my period of time. I'm off again for long periods, but you won't notice that I've gone away for . . .'

Osmond often asked Mayhew to repeat, much as one might ask a maniac the name of the president or the current prime minister to measure their level of derangement, the mystifying phrase 'To be prosperous, a nation requires a safe and secure supply of wood'. The MP didn't manage it very often, and, as time went on (and on, and on) Mayhew was extremely pleased with himself when he occasionally got it right. A more peculiar experience for a first-time user of mescaline would take no little time and ingenuity to devise.

The MP's eyes were alternately alive and entranced, and, even through the hazed-out filter of the black-and-white film, they sparkled in wonder at the beauty of his surroundings, or registered wild, amused bafflement at his inability to add a couple of numbers together. Most charmingly of all, he grinned brilliantly and with the shyness of a small boy startled at play by an adult as reality unravelled and he unsuccessfully attempted to subtract three from one hundred. In later stages of the interview, as Mayhew addressed the camera, his pupils flat, eerily blank discs of infinite blackness, he unnervingly lurched from sense to mystified incoherence and back, as he discussed time and space. 'There is no absolute time, no absolute space, it is simply what we impose on outside space,' he intoned, right on the edge of either a psychedelic breakthrough, or his sanity.

It is one of the strangest pieces of television ever made, and, sadly, one never broadcast, though it would later be sampled by Scottish techno band the Shamen. The BBC consulted a selection of priests, philosophers and sundry other mystics and thinkers who rejected as invalid the experiences of blissful eternity that Mayhew reported. Mayhew himself offered a brilliant deconstruction of their logical fallacy and the primacy of lived experience when he reflected upon the experiment later in his life in the documentary *LSD: The Beyond Within*: 'The psychiatrists afterwards, and common sense, they all said: "This is nonsense. You couldn't have had these experiences [of time expanding to eternity]. There was no time, as the film shows, there was no time for you to have them in." And the psychiatrists would speak and I accept this, they would say I was simply showing the symptoms of what they call the disintegration of the ego, and I accept that, too. At the same time, they didn't have the experience.'

The Mayhew experiment was among the first live 'trip reports' ever undertaken, where a user of a psychoactive compound documents his experiences for the benefit – whether entertainment or education – of others. No lab will suffice, no brain tissues in culture; the mind is the test tube where the reaction takes place, and this experience would inform many explorers that followed him. The fact that this outlandish mental expedition was embarked upon by a distinguished British politician and an internationally respected scientist reveals much about the journey away from a more liberal British stance towards the dysfunctional model of drug prohibition that now prevails worldwide.

LSD remained legal for more than twenty years after its creation, since its users tended to be psychiatrists, scientists and, in the main, other serious-minded researchers. LSD's early users included James D. Watson and Francis Crick, who cracked the fundamental secret of life in March 1953 when they imagined the double helix form of DNA while under the influence of a small dose of the drug, have shaped society in ways unimaginable before its appearance.

Once it escaped the psychiatry ward and other medical institutions in the 1940s and 1950s, LSD was the first compound to enable mass drug use in the West during the 1960s. The mescaline eaten by Ellis was organic, natural and derived from peyote. Huxley, Osmond and Mayhew's hydrochloride was lab-made, but synthetic – and the dose was 400 mg. LSD was active at just 100 μg (micrograms). One small lab could produce enough LSD to dose millions of people. In America, the drug plotted a course from the clinic to the street, just as it would in the UK. Author Ken Kesey, and Harvard professors Timothy Leary and Richard Alpert were among those who evangelized about the substance, believing its use would herald a new age of human consciousness. Owsley Stanley, the world's most exacting and prolific LSD chef who supplied the majority of America's West Coast with LSD in the 1960s, claimed he made so much acid not because he wanted to change the world, but rather because it was almost impossible not to make vast quantities of the drug once the synthesis had been embarked upon. This was also the moment when synthetic, laboratory-made drugs replaced plant-based compounds as the substances most commonly used recreationally, with the exception of marijuana.

Marijuana was in many ways the ultimate gateway drug for both users and lawmakers in the 1960s. Earlier in the twentieth century it had been the drug most widely used by the new black urban American underclass and, later, the mainly white beatnik counterculture. Laws forbidding its use in the US had originally been inspired by an influx of Mexican immigrants at the turn of the twentieth century, who smoked marijuana recreationally. But the enthusiastic uptake of the drug by white, educated, middle-class American youths and students, and its association with the Vietnam- and draft-rejecting hippy counterculture, helped produce the Establishment conviction that drug-taking was a profound threat to society.

The streets and airwaves were flooded with the sights and sounds of psychedelia through the mid-to-late 1960s. In 1965, the Beatles slyly snuck a reference to their first LSD experience into their number one record, 'Help!', in the curious lyric: 'Now I find I've changed my mind and opened up the doors'. Most of the UK had no idea what the drug was, or that it even existed, or that the Beatles had taken it. Neither did the Beatles, initially, since their drinks were spiked with it by their dentist after dinner one evening.[9] But the drug soon entered the culture and it popularized, if not normalized, recreational drug use in that era and beyond, and revolutionized youth culture, sexual politics, music and art on both sides of the Atlantic. Times were changing; in Western Europe and the US drugs saturated everyday life.

LSD was banned in the UK in 1966. During the debate on legislating against the drug, even Britain's staid law lords revealed themselves to be strangely fascinated by it, with one, Lord Saltoun, asking, 'May I ask the noble Lord whether LSD-25 is the drug that enables you to remember what happened when you were born?[10] 'He was pithily answered by Lord Stonham, 'My Lords, I think that the hallucinatory effect created is not to enable you to remember back like that, but rather to forget and imagine that you are otherwise and elsewhere than you in fact are. But, of course, LSD is not the only substance that can create that illusion: I have known people who thought they could fly on four pints of bitter.' The Lords did not offer any evidence-based reasons for banning the drug; they merely mentioned newspaper reports of people jumping from buildings or into lakes after taking it, and a *British Medical Journal* editorial that said the drug should be banned just as amphetamine was.

LSD, mescaline and psilocybin were all banned that year, though there was little scientific proof of their physical harmfulness – bitter beer, meanwhile, remained legal. LSD was banned federally in the US in 1968 after the passage of the Staggers-Dodd Bill, which amended the Food, Drug, and Cosmetic Act.

In the 1960s, another synthetic drug had become popular in a more working-class context: speed. Amphetamines were the first drug other than alcohol and nicotine to be abused recreationally on a wide scale in the UK.

Through the 1930s and 1940s, amphetamines had been widely prescribed by doctors for fatigue, a fact that might go some way to explaining the gruesomely chipper and relentlessly chatty representation of the British squaddie in films of the era – over seventy million amphetamine tablets were used by British soldiers during the Second World War. Popular too with housewives looking for a break from drudgery or to lose weight, they were seldom used recreationally at this time, and supplies were either stolen from factories or diverted from legitimate prescriptions.

The mod cult, born in working-class districts around industrial centres, gained a foothold as conscription or National Service ended in 1960. These dapper working-class kids rejected post-war austerity for conspicuous consumption, their sharply tailored clothes and foreign motorcycles bought on credit. They forged what was to become a classic youth culture template: sexual freedom, new drugs, all-night dancing to music with alien, possibly African rhythms, and a wholesale and unwhole-some disrespect for authority. They rejected booze as a drug for the old, the square, the badly dressed and those unable to dance stylishly, and presented the media with their first ready-made post-war drugs scandal. The UK government swiftly moved to ban amphetamines in 1964, with the main effects being a rapid increase in their use, and higher prices following a twenty-five per cent reduction in availability.

In the mid-1970s, the drug was popular with dancers in the north of England who found that the metronomic beat of classic American soul rocked harder still when jacked up on these banned pills and powders. Punks loved speed, too, the drug's aggression and stiletto-blade sharpness matching their staccato anti-funk, their jackhammer drums and three-chord rants, the perfect powdered fuck-you to the hippy dream of love, peace and self-indulgence and a stinging gob in the eye of goblin-obsessed prog-rockers. From the late 1970s until the late 1980s, other than a brief and ruinous flirtation with heroin in the latter decade, Britain's drug scene was restricted to this limited chemical palette of uppers and downers, acid and hash, and cocaine for the wealthy.

Cocaine had first become widely popular in America in its powdered form in the 1970s, and was ubiquitous among musicians and the wealthy as the acid daze came to an end. Discos of the era, such as New York's glamorous Studio 54, were packed with prancing celebrities with pristinely powdered noses, while the far funkier Loft, run by DJ David Mancuso, took dancers on an all-night journey into the light in a ritualized, almost ceremonial urban setting where LSD was the favoured intoxicant. For all their differences, the contexts in which these drugs were taken, though, were similar – polysexual, multi-racial, music-driven. The cocaine that powered such venues enriched Colombian narcotrafficker Pablo Escobar by three billion dollars in 1989, and the undoubted glamour of the scene was presumably lost on the thousands of Colombians slain by the smugglers. In the 1980s came crack, the more potent, smokeable form of the drug that has ravaged inner cities, where it found favour for its lower price and harder hit. As harder substances replaced psychedelics, problems associated with drug use grew, including addiction, overdoses and inner-city crime.

Drug laws had by this point been tightening for decades, both domestically and internationally. Ostensibly the legislation was introduced to protect the health of nations, and it is impossible to argue that many of the drugs banned worldwide today are not addictive, harmful or problematic if used to excess. Practically, though, it could be argued that the banning of drugs has been as much to do with notions of morality, and with the need for a sober and compliant workforce, as it has been to do with protection.

International drug law is today controlled by three United Nations treaties: the Single Convention on Narcotic Drugs, 1961, the UN Convention on Psychotropic Substances, 1971 – a draft of which was released in 1969 and served as a blueprint for later American and British drug laws – and the United Nations Convention against Illicit Traffic in Narcotic Drugs and Psychotropic Substances, 1988, which bans commonly used chemicals required to produce and synthesize illegal drugs. The vast majority of UN member states are signatories to these three conventions, and bound by law to criminalize the production, distribution, purchase, sale and possession of the hundreds of drugs they list for anything other than scientific or medical purposes.

In the UK, drugs are currently controlled by two Acts of Parliament, the 1968 Medicines Act and the 1971 Misuse of Drugs Act. It is the latter act that is usually cited in criminal drugs cases brought against individuals, and it prohibits the production, supply and use of hundreds of individually named compounds. It separates drugs into three classes, C, B and A, with progressively harsher punishments for those involved with drugs that are deemed to be more dangerous.

The US, a key driver in the internationalization of drug controls, consolidated decades of piecemeal legislation into the 1970 Comprehensive Drug Abuse Prevention and Control Act. It named specific drugs that were banned federally, and classified drugs into five different bands, or schedules, based upon their addiction potential, with corresponding punishments for each. On 17 June 1971, serpentine president Richard Nixon announced the start of formal hostilities as he fired the first shot of the War on Drugs: 'Ladies and gentlemen: I would like to summarize for you the meeting that I have just had with the bipartisan leaders which began at eight o'clock and was completed two hours later. I began the meeting by making this statement, which I think needs to be made to the nation: America's public enemy number one in the United States is drug abuse. In order to fight and defeat this enemy, it is necessary to wage a new, all-out offensive.'[11]

The US Drug Enforcement Agency (DEA) was established in 1973, with responsibility for enforcement of drug policy handed to the Justice Department rather than the Treasury, underlining the new political view of drug abuse: previously an economic crime of tax evasion, now it was framed as a socio-moral offence.

In a pattern that repeats to this day, drug laws actually exacerbated the social problems associated with drugs. It was the emergence of synthetic drugs from the 1960s onwards that enabled the creation of a mass market and a recreational drug counterculture, and that in turn was powered by the alchemical ability of chemists to produce drugs using nothing more than laboratory equipment and chemicals. But it was the act of banning substances for which demand was so high that made their manufacture, import and supply so disproportionately profitable – and popular. Drug laws have produced a situation where an ounce of gold, at around £1,000 in late 2012, costs about the same as an ounce of cocaine of average purity, which is a simple extract from a plant that grows wild with little attention. A kilo of crystal meth can be made for a few hundred dollars in Mexico, and is sold on the streets in individual deals for hundreds of thousands of dollars.

The process by which drugs are banned in the UK has been guided, since 1971, by the Advisory Council on the Misuse of Drugs (ACMD). It is a twenty-strong committee with experts taken from many disciplines – pharmacology, toxicology, neurology, criminology, law, chemistry and many more – whose job is to assess a drug's harms via a process that draws on their collective expertise. This involves dozens of scientific procedures, and it can take many months to come to their conclusions; when they do, they recommend the best course of government action. It is not currently possible under British law simply to ban a compound because it has a psychoactive effect.

Even in the first few years after the 1971 Misuse of Drugs Act came into force, rogue chemists were pushing at its boundaries. Laws around drug use and drug chemistry tend to be highly complex, and suppliers will always find ways to sidestep the legislation.

In 1975, a chemist in the Midlands had been churning out ring substitutions on illegal drugs, making hallucinogens and stimulants that did not feature in the Act. Specifically, he had been making a designer drug named bromo-STP, a potent hallucinogen. Police had found this and other new designer drugs on the streets, and the ACMD had recommended that it be banned. Britain uses a generic model of drug categorization and control, and so the government asked the ACMD to look at the drugs the Midlands chemist was making and to extrapolate from there the likely steps he or others might take next, using their collected chemical knowledge to suggest a law that might pre-empt such steps.

John Ramsey, chief toxicologist at St George's, University of London, explains the process. 'When the government is trying to write a new drug law, advisors to the legislators are asked to try and guess what other compounds might be made by chemists in order to get around the legislation they are trying to put in pace. If you didn't do that, it'd be like having a speed limit just for blue cars, when you want a speed limit for all cars,' he says.

In the UK in 1977, the House of Lords met to debate these pro-posed changes to the Act, which sought to outlaw many dozens of variations on the basic tryptamine and phenethylamine structures. 'The Council took note of the fact that further substances of a like nature but which would not be caught by the entries in the Schedule to the Act could be synthesized with relative simplicity by making minor molecular changes to the basic structure of the chemical,' said Lord Wells-Pestell on 20 June 1977 in the House of Lords. 'The Council accordingly instructed its Technical Sub-Committee to explore the possibility of closing the door on the production of such substances for misuse purposes, but with the minimum interference with possible legitimate medical and research use, by controlling all possible variations in each series by means of a generic description, or formula.'[12]

He then asked his learned friends to forgive him for what he was about to say next. He was attempting to communicate major and complex new proposed changes to Britain's drug laws. The chemical relay race was about to begin in earnest, with lawmakers attempting to keep up with a discipline – or a crime – they barely understood and which, even then, was quicker and more responsive than the legislative process. 'The Technical Sub-Committee found that the particular compounds which produced undesirable hallucinogenic effects fell into two categories. First, compounds derived from tryptamine, or from ring-hydroxy tryptamines which have been substituted at the nitrogen atom of the side chain by an alkyl group or groups. Examples of this category which are already controlled under the Act are psilocin and psilocybin. The second category was that of compounds derived from phenethylamine by alkyl or other substitution in the aromatic ring. Examples of this category which are already controlled under the Act are mescaline, and STP and Bromo STP to which I have already referred.'

Lord Platt concurred, and thanked Wells Pestell for his work, and spoke for everyone present when he confessed his mystification.

'My Lords, it is customary for the layman to associate the doctor with some supernatural powers. I should like to assure noble Lords that there is not one doctor in thousands who understands the exact nature of these drugs. The noble Lord has made clear to us the reasons for this order and the kinds of substances which are included in it. I should like to support it.'

This judgment led to a ban on drugs such as MDMA in the UK long before they ever became available here, and dozens of other drugs, in a similar fashion, at one stroke. But it did not ban all of the new possibilities. A precedent was set and a cycle began.

Whatever your stance on their efficacy and fairness, Britain's drug laws have historically been subtle, intelligent and complex pieces of legislation that draw on the expertise of many different strands of science. The United States has a different method of legislating around designer drugs, one which bans compounds on the loosely defined basis of their effects and similarity to banned drugs, which is discussed in full in the next chapter.

But subtle or not, all of these laws were written in an era before mass communications – before most homes even had a telephone line or a colour television, when news came a few times a day on screens and twice a day on paper. They were drafted in an age when air freight costs were prohibitively high for individuals, in an age when communication with distant, communist China was so slow as to be impossible. When they were created, computers were room-sized, and were owned in the main by governments. They were first written, that is to say, almost half a century before the web was born. These laws were made five decades before the creation of an entirely new drug whose effect on users would be different from that of LSD, but equally profound. This drug would leak into the global water table on a scale that would have given even the most extreme LSD evangelist pause for thought.

One individual, allied with technology, would be a central figure in this new race between chemists, users, the culture and the law: American Alexander Shulgin, the world's most prolific and genius-tinged psychedelic chemist, the godfather of Ecstasy.

# 2

## The Great Ecstasy of the Toolmaker Shulgin[1]

A squirrel-infested shack containing a chemistry laboratory lies at 1483 Shulgin Road, Lafayette, California 94549. Its grounds are strewn with cacti and fringed with greenhouses; the front door is rickety, its hinges rusted now. This is the unlikely epicentre of a global drugs culture. The products that have emerged from it, the methodology that produced these new compounds and the career of its owner make it, indisputably, the world's most storied and influential drug lab.

For much of the last century Alexander Shulgin worked in relative obscurity. But in the mid-to-late 1980s, a new drug, MDMA, later known as Ecstasy, started appearing on the streets of the USA and Europe. This substance, a stimulant that prompted emotional openness, would change the world's drug habits for ever, bringing the psychedelic experience to millions who, before its advent, would perhaps never have considered using drugs.

Alexander, or 'Sasha', Shulgin is, depending on your viewpoint, one of the greatest and most under-celebrated scientists of the twentieth century, or an irresponsible folk devil who has corrupted millions and killed dozens with the drugs he has created. He says he's just a toolmaker. His most famous tool was MDMA, or 3,4 methylenedioxymethamphetamine; Ecstasy.

Born in 1925 to Russian immigrant parents, Shulgin had an unremarkable early life, dropping out of Harvard aged nineteen to join the Navy. In the 2011 documentary about his life and work, *Dirty Pictures*, he mentions, casually, that for the years he was at sea he decided to memorize an entire biochemistry manual, thinking it would be 'a neat challenge'. Those years of unwavering discipline are curiously at odds with his public image – which he hates – as Dr Ecstasy, a wild-haired, archetypal mad scientist. 'It brings an element of notoriety that does no good,' he has said.

MDMA's path from obscurity to ubiquity is so improbable as to sound fictional. Chemists at the German pharmaceutical company Merck could not have guessed, as they quietly synthesized the world's first batch in 1912, that their work would one day be used first as a tool in psychotherapy – the discipline did not even exist at that time – nor that it would later surface as one of the world's most popular recreational drugs. Referred to in the academic literature as 'methylsafrylamin', the new substance was designed to enable the company to make a clotting agent without having to produce it via a patented intermediary, or, to quote the firm's papers, it was 'a precursor in a new chemical pathway which was patented in order to avoid an existing patent for the synthesis of [a] clotting agent, hydrastinine'.[2]

The instructions for the synthesis – the chemical construction in the laboratory – of MDMA lay undisturbed in the Merck archives for decades. But the foundations for much of today's chaotic international drug scene were unwittingly laid in that very laboratory. Shulgin resynthesized MDMA in 1965, after receiving a mysterious tip-off about the compound from a fellow researcher in the field of psychedelic drug synthesis. He has never revealed more than that.

MDMA is just one of the hundreds of psychedelic agents Shulgin synthesized and then tested – by ingesting them. After noting his experiences under each drug's influence, he would carefully document his observations. Like a chemical Noah, he then carefully ushered his creations into safety, his chosen arks a pair of printed books containing dense and technical instructions on how to make them. His influence on generations of drug users is immeasurable yet mainly unknown; and his influence on the future of drug taking is only just starting to be correctly understood.

In common with Huxley, Mayhew and Osmond, Shulgin's chemical career began after a dose of mescaline hydrochloride crystals in the early 1950s. He spoke of the intensity of colours he experienced, of a childlike wonder, of accessing long-buried memories: 'I thought: "What it's doing is allowing me to communicate with parts of me that I had not communicated with for a long time."' He was, he says, astonished that a small quantity of white powder had unearthed such memories. If the powder did not itself contain those memories, he reasoned, then the answer to the mystery must reside within his own mind: 'I understood that our entire universe is contained in the mind and the spirit. We may choose not to find access to it, we may even deny its existence, but it is indeed there inside us, and there are chemicals that can catalyse its availability.'[3]

In 1954, Shulgin returned to his education, this time at the University of California, Berkeley, where he earned a PhD in biochemistry. He then worked for the Dow chemical company, where in 1962 he invented a biodegradable insecticide named Zectran. The product was so profitable that Dow gave him free rein to produce any compound he desired. He chose psychedelics, and set to work with the zeal of the convert, exploring all the possible variants of the mescaline form that had fascinated him so much.

His experiments were carried out using precursors, or chemical building blocks, and reagents that set off chemical reactions to produce hundreds of new drugs, compounds that, until he invented them, did not exist anywhere in the universe. The process of producing new drugs in this way is known as ring substitution.

The connections between basic chemical structures, drugs and the brain's neurotransmitters can be represented in a molecular triptych. First, here is the molecule for phenethylamine:

*Phenethylamine*

Starting from the left of the molecule, the hexagonal structure is known as a phenyl group, and each of its vertices represents a carbon atom. This illustration shows a hexagonal ring of six carbon atoms. If that ring were not connected to any other groups or molecules, it would be known as a benzene ring. A hydrogen atom bonds to each carbon in that ring, but for visual economy, it is not represented here. Bolted on to the right of the phenyl group is an ethyl group, consisting of two more carbons (each of which is also bonded to more hydrogens). Finally, the last structure to the right represents the amine group, which is made up of one nitrogen and two hydrogen atoms connected to a carbon.

'Phen-ethyl-amine' is naturally produced in our bodies and brains, especially when we first fall in love, and there are high levels of it in chocolate, cheese and sausages. It is a neurotransmitter – a molecule that passes around our brains and activates receptors that ultimately govern whether we are hungry, angry, elated or sleepy. Phenethylamine also modulates other neurotransmitters and can accentuate their effects upon our minds and bodies.

Second, here is a neurotransmitter that is related to and resembles phenethylamine, dopamine:

*Dopamine*

Dopamine is normally pumped around our brains when we are excited or sexually aroused, taking risks, or seeking or receiving rewards – or taking any one of dozens of other stimulants. It regulates muscle movement, the sensation of pleasure, and motivation.

And here is the third molecule in our triptych, mescaline:

*Mescaline*

The same basic phenethylamine skeleton is present in the second two molecules, but we can see that there are a number of other molecular groups attached to it. (These diagrams are, incidentally, 2D renderings of 3D objects.) The mescaline molecule fits into the dopamine receptors in the brain, and, in some way yet to be understood, affects our perception of reality in radical and unusual ways.

Shulgin took the mescaline molecule, and swapped the hydrogen molecules attached to each of the carbons on the benzene ring for other chemicals. He added sulphur molecules, or cleaved off an atom or two of oxygen here, spliced in an iodine or a bromine there, exploring the possibilities of chemistry.

Shulgin's intense, rigorously academic study gave him pointers towards the likely effect and dosages of most of the new drugs he was inventing. But sometimes he simply added a chemical at various different positions on the ring and tested the drugs on himself to see what happened. He carried out the same process with both the mescaline molecule and with variants on the tryptamine skeleton, the parent structure for magic mushroom-like drugs.

Where mescaline most closely resembles the phenethylamine and dopamine neurotransmitters, tryptamine resembles serotonin. Drugs containing tryptamine work by entering serotonin receptor sites in the brain, and by some sub-atomic magic, triggering a release of it, which affects mood, social behaviour and impulsiveness. It limits other neural activities and changes our perception of reality in the process. How these substances achieve this in neurological terms is hotly debated, but the latest research suggests that when flooded with serotonin the mind dampens down brain activity, rather than increasing it, as had long been thought. (These descriptions deliberately and necessarily simplify the interaction between drugs and the brain, since each category of drugs actually acts on both systems at once, and each system is far too complicated for any non-scientist to describe – or read about.)

As the social taboos and moral panics around psychedelics started to build, Dow found itself the uneasy holder of patents for powerful psychedelics such as DOM – a designer drug created by Shulgin and popularized following the outlawing of LSD. This potent drug flooded the streets of Haight-Ashbury in 1967 and is said to have caused many traumatic episodes as tablets containing guaranteed overdoses of the chemical circulated in their thousands. Syd Barrett of Pink Floyd is said to have been profoundly affected by the drug in the years preceding his nervous breakdown and withdrawal from society.[4]

Shulgin left Dow in 1965 and set up his home laboratory in Lafayette, where he continued engineering new drugs from the basic mescaline phenethylamine structure. In the process he revolutionized our knowledge of biopharmacology, and set in motion, unwittingly, the global shift in drug use that has been witnessed since the mid-to-late 1980s. He worked for the DEA as an expert witness, supplying them with reference samples for use in criminal trials, and was otherwise left to his own creative devices.

Shulgin blurred the lines between the epiphanies of art and the harder edges of science. His career during this period can be seen, by those who are sympathetic, as a reinvention of the American pioneer spirit. He pushed the boundaries of his own mind, searching for a manifest destiny of his own, for an American expansionism anyone could believe in – but with no harm to others ever intended, and no presumption of ownership ever made. With his unruly beard, ready grin, colourful baggy shirts and loping gait, the endlessly punning chemist worked in his ramshackle garden shed laboratory, surrounded by thickets of glassware and dangling, vine-like tubing, the air thick with volatile gases and the thunderous chords of his beloved Rachmaninov.

Eventually, sensing a change in the attitude of the DEA towards his work, Shulgin published his entire body of know-ledge in two seminal works of psychedelic chemistry: *PIHKAL* (*Phenethylamines I Have Known and Loved*): *A Chemical Love Story*, in 1990 and, seven years later, *TIHKAL (Tryptamines I Have Known and Loved)*, which covered the other major class of psychedelics. Psychedelic pioneer Timothy Leary has compared *PIHKAL* to Charles Darwin's *On the Origin of Species*,[5] and, while most of Leary's grandiloquence can be safely disregarded, it's arguable that on this he was right. The DEA did not share that view, and they raided Shulgin's lab in 1994, accusing him of possession of compounds he was unauthorized to own. Shulgin was found to be in possession of various samples of drugs people had sent him, anonymously, to test, which was an infraction of his DEA licence.

*PIHKAL* reveals in practical detail the chemical synthesis and human dosage of hundreds of psychoactive substances, each of which are in the phenethylamine class. At the time of its release, many of the precursor chemicals, that is, the ingredients required to produce them, were not banned and were available without excessive governmental interference.

While it's in no way comparable in terms of complexity and difficulty, drug chemistry is, to all intents and purposes, much like baking. Both aim to take one type of matter and use the physical forces of heat and chemical reactions to transform it. The cook and the chemist both need ingredients, techniques, equipment and expertise. Where a cook uses flour and fat and heat, a chemist such as Shulgin might use an essential oil, such as oil of parsley, or safrole, as a precursor, its similarity to pre-existing drug structures, such as amphetamine, making fewer chemical reactions necessary.

Indeed, in *PIHKAL* Shulgin identified ten amphetamines that were closely linked to essential oils and he deliberately chose precursors that were, back then, available without a great deal of trouble. 'What are these relationships between the essential oils and the amphetamines?' he wrote. 'In a word, there are some ten essential oils that have a three-carbon chain, and each lacks only a molecule of ammonia to become an amphetamine.'[6]

Since their publication *PIHKAL* and *TIHKAL* have served as guidebooks to the new chemical terrain, their pages revealing in detail the doses of new and unknown drugs. But these are not orthodox academic manuals. The books mix chemistry and synthesis guides with wry asides, trip reports, and the story of the love affair between Shulgin and his wife, Ann, a counsellor and psychotherapist who has used many of her husband's drugs in her practice, with, she says, great success. *PIHKAL* also contains an extended riff on the reasons behind Shulgin's quest for psychedelic experiences. These drugs were for research, for exploration, and for the pursuit of spiritual and psychological insights, he writes:

> I am completely convinced that there is a wealth of information built into us, with miles of intuitive knowledge tucked away in the genetic material of every one of our cells. Something akin to a library containing uncountable reference volumes, but without any obvious route of entry. And, without some means of access, there is no way to even begin to guess at the extent and quality of what is there. The psychedelic drugs allow exploration of this interior world, and insights into its nature.

Shulgin also reflects upon the unprecedented nature of the attacks made by governments upon those whom he sees as seekers of spiritual truths:

> Our generation is the first, ever, to have made the search for self-awareness a crime, if it is done with the use of plants or chemical compounds as the means of opening the psychic doors. But the urge to become aware is always present, and it increases in intensity as one grows older [. . .] This is the search that has been a part of human life from the very first moments of consciousness. The knowledge of his own mortality, knowledge which places him apart from his fellow animals, is what gives Man the right, the license, to explore the nature of his own soul and spirit, to discover what he can about the components of the human psyche.[7]

Shulgin argues passionately for free speech and the rights of the individual; there is no one less un-American than this chemist savant, yet he remains an outcast even today, in his old age. He asks: 'One of the major strengths of our country has been in its traditional respect for the individual . . . How is it, then, that the leaders of our society have seen fit to try to eliminate this one very important means of learning and self-discovery . . .?'[8]

Luckily for science, Shulgin and his friends decided that their minds and bodies were their own dominion, and self-experimented with the new compounds he created. Shulgin would gradually increase his doses by minute increments until the desired effects were felt. He would report on the drugs' efficacy and their impact on what he called 'the erotic' – he and his wife agreed that any drug that precluded their lovemaking was undesirable – and his research group would offer their findings on the drugs' effects on music appreciation and their artistic sensibilities in genteel and discreet prose. Shulgin even pioneered the concept of the 'museum dose' – a quantity of a drug that he deemed appropriate for trips to interesting exhibitions. He and his group would measure each of the experiences according to an agreed scale and write up their impressions, from a 'plus one' up to a 'plus four'. The 'plus four', he wrote in *PIHKAL*, was a 'a rare and precious transcendental state, which has been called a "peak experience", a "religious experience", "divine transformation" a "state of Samadhi" and many other names in other cultures.'[9]

Each entry begins with a densely complex synthesis for each drug. A typical entry, for TMA, a mescaline derivative, reads thus:

To a solution of 39.2 g 3,4,5-trimethoxybenzaldehyde in 30 mL warm EtOH there was added 15.7 g nitroethane followed by 1.5 mL n-butylamine. The reaction mixture was allowed to stand at 40°C for 7 days. With cooling and scratching, fine yellow needles were obtained which, after removal by filtration and air drying, weighed 48 g. Recrystallization from EtOH gave 2-nitro-1-(3,4,5-trimethoxyphenyl) propene as yellow crystals with a mp of 94–95°C.[10]

The remaining 600 words of chemical nomenclature and technique would make even less sense to anyone except highly trained organic chemists. But then the register shifts dramatically as users describe their experiences:

> [With 225 mg of TMA] There was quite a bit of nausea in the first hour. Then I found myself becoming emotionally quite volatile, sometimes gentle and peaceful, sometimes irritable and pugnacious. It was a day to be connected in one way or another with music. I was reading Bernstein's *Joy of Music* and every phrase was audible to me. On the radio, Rachmaninov's Second Piano Concerto put me in an eyes-closed foetal position and I was totally involved with the structure of the music. I was suspended, inverted, held by fine filigreed strands of the music which had been woven from the arpeggios and knotted with the chords. The commercials that followed were irritating, and the next piece, *Slaughter on Fifth Avenue*, made me quite violent. I was told that I had a, 'Don't cross me if you know what is good for you' look to me. I easily crushed a rose, although it had been a thing of beauty.

Of the hundreds of phenethylamines he explored or invented Shulgin identifed six as being particularly useful for the exploration of inner space, and for understanding reality in new and extraordinary ways. These were mescaline, DOM, 2C-B, 2C-E, 2C-T-2, 2C-T-7. The differences between each drug can be measured, prosaically, by descriptions of their dose, their duration and their chemistry. More poetically, members of his test group would discuss them as wine buffs might discuss a fine Pétrus, appreciating the finer points of their effects. Some approached it from a spiritual perspective, and ranked the drugs on the insights the compounds gave them into their own natures.

The phenethylamine that would catapult Shulgin's work from academic and even psychedelic obscurity into the mainstream, though, would be MDMA, or Ecstasy. MDMA does not exist anywhere on earth naturally, but its synthesis can be arrived at in a couple of days by a moderately competent chemist, given access to the right materials and equipment. There are a number of different routes to the production of the drug, including isosafrole, MDP-2-P, piperonal and beta-nitroisosafrole. The easiest synthesis for MDMA, though, is from an essential oil, safrole, that acts as a natural pesticide for the plants that produce it. With a sickly-sweet synthetic smell somewhere between marzipan and aniseed, the oil is used legitimately in the perfume and flavourings trades; but buying it in any quantity will ensure your almost immediate presence on a police watch list, as it is a well-known and internationally banned precursor to MDMA.

In around 1965, Shulgin synthesized the drug using safrole as a precursor, and in 1967 he started to work with the new drug himself. Entry 109 in *PIHKAL* would prove to be the most revolutionary of the hundreds it contained. *PIHKAL* does not attribute authorship to its substance reports, and Shulgin himself never (publicly) rhapsodized about the power of MDMA, comparing it instead to a low-calorie martini. One of his test group, however, felt somewhat differently:

(with 120 mg) I feel absolutely clean inside, and there is nothing but pure euphoria. I have never felt so great, or believed this to be possible. The cleanliness, clarity, and marvellous feeling of solid inner strength continued throughout the rest of the day, and evening, and through the next day. I am overcome by the profundity of the experience . . .

(with 120 mg) The woodpile is so beautiful, about all the joy and beauty that I can stand. I am afraid to turn around and face the mountains, for fear they will over-power me. But I did look, and I am astounded. Everyone must get to experience a profound state like this. I feel totally peaceful. I have lived all my life to get here, and I feel I have come home. I am complete.[11]

What exactly did these molecules do to the human mind? What is the chemical process by which a certain arrangement of atoms can make a user experience bliss? How can a chemical make an instant seem eternal, or transform the banal into the transcendent, or turn a blinding spotlight on to ineffable universal mysteries, for a brief, possibly delusional moment?

Drugs hijack our neural circuitry and change the way we feel, how we see and hear and interpret the world, and in the case of some psychedelics, how we see ourselves. As we saw above, they work in this way, in the most basic terms, because their molecular structures closely resemble naturally occurring chemicals, or neurotransmitters, that are generated by our brains to regulate mood, muscle action, appetite and dozens of other important physical processes. Drugs act as chemical 'keys' that somehow fit into the transmitters' 'locks' in our brains. These molecules circulate, and are received back into the site that transmitted them, much like hot water in a radiator and boiler system.

Prozac, the antidepressant, works on the brain's levels of serotonin, the chemical that is triggered when we experience stimulus that results in joy, but the effects of this and the dozens of other pharmaceuticals in its class are gradual. They work more on inhibiting the synapses' reuptake of naturally circulating serotonin, rather than triggering unusually large quantities of it.

When we take MDMA, serotonin floods the brain, and our synapses' regular pattern of release and reuptake are dramatically changed by the presence of the drug.

There are also serotonin receptors in the heart and the stomach, which goes some way to explaining the concept of gut instinct, or feeling lovesick when missing a partner, the sensation of butterflies when meeting a loved one after an absence, or the real and physical trauma of a truly broken heart.

MDMA also produces a flood of dopamine, the neurotransmitter pumped around our brains when we are excited, or sexually aroused, or seeking or receiving rewards. And finally it also leads to the release of norephrenine, also known as noradrenaline, a neurotransmitter released when we are frightened or angry or edgy, which increases respiration and heart rate.

The magical ratio of serotonin-dopamine-norephrenine release and reuptake prompted by the presence of MDMA in the human brain is not produced by any other substance. Other recreational drugs come close, but the charged, empathetic, transcendent, peak-MDMA experience has yet to be matched by any designer drug. It's called Ecstasy for a reason.

This fascinating, mysterious collision between the physical and the intangible, between the flesh and photon, reframes all human consciousness as a chemical reaction. That's the science of it, but human emotion and the mystical experiences some drugs afford users cannot be so neatly quantified.

MDMA was used initially in the 1970s by psychotherapists in the USA excited by the drug's therapeutic potential. One such was Leo Zeff, who had been introduced to the drug by Shulgin in 1977. The California-born practitioner was so thunderstruck after his first experience that he cancelled his retirement and travelled the country dosing thousands of willing users with the drug, whose extraordinary qualities of personal boundary dissolution he believed would help therapists to discover the root of a sufferer's problems. Zeff claimed that a single MDMA-assisted therapy session could accomplish as much as six or more months of traditional therapy. Therapists say the drug first helps patients to open up and discuss their problems honestly, and then enables them to engage in objective analysis without self-judgement, and finally allows them to generate solutions.

Shulgin hated the term Ecstasy, preferring the stark chemical nomenclature of MDMA. The name Ecstasy was popularized by an ex-priest-turned-MDMA-propagandist, Michael Clegg, whose E-triggered epiphany was so utterly damascene when it came that he started to press the pills himself and give them away for free. But this was the yuppie era, and by 1984 the 'Texas group' – a loose conglomeration of dealers – was selling 500,000 pills a month in that state alone. When he was warned that if he carried on selling it so openly the drug would be banned, Clegg is said to have responded: 'Yes, and then I'm going to get very rich.'[12]

In the mid-1980s, designer Phillipe Starck's eponymous Dallas, Texas, nightclub was a hotbed of early public, hedonistic MDMA use, a tactile bacchanal with unisex bathrooms, playing kitschy synthpop and frequented by patrons with prominent cheekbones and heavily moussed hair. The pursed lips and swinging hips, the primped pomp and sexual display, the high-sheen finish and conspicuous consumption could not have been more removed from a therapist's sofa: it was an orgiastic, consumerist 1980s riot with guest star appearances by imperious Jamaican diva Grace Jones. Fittingly, a CD compilation of the music played at the Starck club sold in 2007 for over US$14,000 on eBay.

Like a dandelion in the wind, the seeds of Ecstasy were blown randomly out from these small pockets of enthusiastic early adoption, finding root wherever they landed. With each blooming came fresh converts who themselves became evangelists for the new drug.

Some say the drug first crossed the Atlantic in the pockets of the sannyasin, or followers of Bhagwan Shree Rajneesh. The orange-robed ascetics, whose guru drove a Rolls-Royce, were evicted from their Oregon commune in 1984. Among Rajneesh's followers, drawn from many New Age disciplines, were psychotherapists who had encountered the drug in the therapeutic communities. A former bodyguard to the guru, Hugh Milne, has claimed rich donors were spiked with the drug – perhaps explaining their beatific view of the charlatan, and their altruistic responses to his requests for money.[13]

In 1984, some of the sannyasin were involved in one of the world's most serious bio-terror attacks, when devotees laced food in salad bars with salmonella bacteria in an attempt to incapacitate voters in the Wasco County elections, in the hope that a lower turnout for the opposition would enable them to install their own candidates. Over 700 people became sick, though none died. The sannyasin left and some of them landed on the hippy island of Ibiza, the freak-zone safety net for Franco-era Spain, a blinding white Balearic haven that later attracted a ragtag gang of refugees – hippies, artists, liberals, discontents, leftists, gay men and women, and the rich at play.

'Sannyasins enjoyed parties, participated heavily in the nightclub life, and introduced various New Age techniques for self-development brought from the USA, including the use of MDMA for meditation and body-therapies,' wrote author and academic Anthony D'Andrea in the island's *Cultura* magazine of summer 2001.[14] The drug soon found its way onto the dance-floors, and into the mouths and serotonin receptors of the wealthy, the famous and the bohemian.

Journalist Peter Nasmyth's series of articles for style and fashion magazine *The Face* in 1986 discussed the pharmacology of the drug,[15] but even at this stage, it is revealing that the interviewees included progressive British psychologist R. D. Laing and name-less pop stars rather than everyday users – for there were almost none. But from 1988 onwards, young British men and women would change that.

The drug was already illegal in Britain at that time, added to the Misuse of Drugs Act in a Modification Order in 1977 after a clandestine chemist had been caught emulating Shulgin's efforts. The Modification Order cleverly specified the mechanisms of ring substitution – molecular replacements of the kind that Shulgin was carrying out – that would make each related compound illegal. This banned most, but not all, of *PIHKAL* in the UK before it was even published.

In the summer of 1988, the drug was discovered by a group of English clubbers while they danced to the hotch potch of Balearic sounds spun by DJ Alfredo Fiorito in the open-roofed, palm-tree strewn Amnesia nightclub in Ibiza. These suburban Londoners brought the culture, the attitude and news of the new wonderdrug to the UK. Supplies followed soon after, produced by Dutch, Belgian and American chemists, and smuggled in by organized crime gangs from Manchester, London and Liverpool.

When Ecstasy first started to be used in significant quantities in the UK, it felt like being a member of an amusing cult. You could spot fellow members on most high streets not just by their floppy fringes and ponytails, their loose-fitting, easy-to-dance-in clothes – their dungarees and Wallabees or Timberlands – but by the unexpected urban eye contact between strangers, the flash of recognition from ten paces. Many British high streets warped into live-action Keith Haring dayglo cityscapes. For one or two brief summers before the utopianist culture was appropriated, gobbled up and spat out like flavourless gum, there was a strange magic at work, as a whole section of the nation's youth seemed, as one, to chill out.

Even the concept of 'chilling out' is an old Ecstasy meme, the phrase borrowed from jazz-era hep-talk originally, and referring to the re-entry period at home after the club with good friends, or the lounging on bean bags in the back rooms of nightclubs by E-heads overwhelmed by the drug, taking time to relax into the rush and cool down, as psychedelic jazz and other layered sample madness careened around the speakers.

For the uninitiated to understand the singular love felt by some British drug users for MDMA, or Ecstasy, there's not a lot they can do except take it. Preferably, they would dose just as they jumped out of a time machine into late-1980s Britain where the nation's youth self-medicated, rejecting the strictures of a drab, individualist puritanism, Thatcherism's dreary, market-obsessed world view, unemployment, heroin, nuclear paranoia, a dirge-like indie-rock alternative scene, and a pop chart dominated by trite, brilliantly bubblegum pop, choosing instead the intense collective euphoria of Acid House.

The emergence of a global dance and drug culture in the late 1980s and early 1990s has left us with what could be argued to be the principal model of illegal youth culture in most of the world. Every town now has a nightclub where you'll find people taking Ecstasy and dancing, from Reykjavik to Buenos Aires, and Bangkok to Moscow. True, its popularity has waxed and waned along with the quality of the product on offer, but MDMA is now undeniably an integral element of all late-night entertainment venues playing music to dance to, at festivals, parties, pubs or prison cells.

Use increased on a hockey-stick graph from 1988 onwards, partly because the prurient and sexualized reporting about the drug delivered a clear and concise message to most young readers: there was a new drug in town, a new scene, and a new beat, and there was tremendous fun to be had. But it also increased because of the drug's convenient routes of administration: orally, via branded tablets stamped with trustworthy logos, making them look much like regulated pharmaceuticals, which took away the fear associated with the needles and powders of heroin and cocaine, the dangers of disease and addiction.

Crucially, Ecstasy did not cause hallucinations of the kind associated with classical psychedelics, such as LSD and mushrooms. The fear many people associated with psychedelics – of losing control or being unaware of their surroundings – dissipated the first time they felt the benign psychic slam delivered by a dose of Ecstasy. The main thing they wanted to do was dance and hug people. This made the experience accessible and popular with many who would never have considered taking drugs, and guaranteed its proliferation.

Most information about Ecstasy at this time was mediated by newspapers censorious about the new drug. Before the net gained mass appeal in the mid-1990s, one of the earliest and most level-headed sources of information on the chemistry and action of MDMA was Nicholas Saunders's book *E For Ecstasy*, published in 1993. The book offered a calm and information-rich analysis of the history and use of the drug, and opened with the author's account of the first and subsequent times he took the chemical. Saunders was a member of London's squatting counterculture through the 1970s and 1980s, and his self-published 1970 guide to living in Britain's capital at minimal or zero cost, *Alternative London*, was essential reading for the British hippy underground. An alternative entrepreneur, he opened Neal's Yard in Covent Garden, turning derelict warehouses into one of central London's first whole-food shops.

In *E For Ecstasy*, Saunders described his first experiences with the drug in 1988 with the rhapsodic tone that was so commonplace at the time:

> When we got off the train I took deep breaths and the air felt wonderful. It was good to be alive. But the intellectual part of myself asked 'What is different to normal? Why isn't life always like this?' I deduced that I was simply allowing myself to enjoy what had always been there. I realized that I had got into the habit of restraining myself. It was not this drug-induced state that was distorted – it was what I had come to accept as my normal state that was perverse. I then realized that over the past few years I had been mildly depressed. And, what's more, I could see why:

some years before I had felt cheated in a business deal, and had carried a resentment like a burden ever since: instead of hurting the person involved, I had been grimly taking it out on myself. This realization and the experience of a few hours 'freedom' was just the tonic I needed; I let go of the resentment and started afresh with new enthusiasm.[16]

After his experience, this driven man gathered every scrap of research he could find on the drug and dedicated the years until his death in a car accident in South Africa a decade later to preaching a message of harm reduction, and telling people how to use drugs more safely.

For his book, Saunders interviewed a wide range of figures about their experiences of taking the drug, including a rabbi, a Rinzai Zen monk, a Soto Zen monk and a Benedictine monk who said the drug gave him similar effects to some of his most profound religious experiences. Saunders also detailed the drug's darker, more negative sides in an honest appraisal that was sorely lacking in mainstream coverage. As you expect with a prohibited drug, supplies at that time were not always pure or safe, but Saunders spread the news of contaminated pills as fast as he could, in the era when words and information were bound by the gravity and solidity of printed matter.

In 1992, a huge batch of pills known as Snowballs came into mass circulation in Europe. They had been synthesized in Latvia, an Eastern European country emerging from communism, in a bid to generate some foreign currency. But instead of MDMA, the pills contained a similar, equally illegal outpost in the methylenedioxyamphetamine family tree, MDA. That year, the music mutated away from the hypnotic house sounds that had characterized the previous years and sped up, becoming darker and more fractured, the clattering breakbeats of hardcore tumbling across cartoony vocals, as sub bass boomed through ever more improbably huge sound systems. Europe was awash with these strange pills that made the atmosphere and users edgier. The once-peaceful rituals of the E-generation started to vibrate uneasily with an outlaw, renegade menace that was reflected in harsher policing.

The lab that synthesized the Snowballs also unleashed a wave of extraordinary hallucinations across the UK and Europe that year, since the chemists had inaccurately dosed the pills with almost 200 mg of the active substance, which is generally enjoyed at 100 mg or below. Many clubbers at that time reported hitherto unseen phenomena on the dancefloor, such as the sudden invasion of thousands of masked, bearded or bespectacled dancers; ornate baroque sofas disappearing into puffs of dried ice and glitterball trails; ancient hieroglyphics swarming across the walls and faces all around them; indecipherable Mayan technological artefacts materializing for an all-too-brief instant before reconstituting themselves as cigarette machines; antic, merrie maypole scenes; dancing Madonnas beckoning seductively with laserbeam hands as the room exploded in rhythm – plus an awful lot of projectile vomiting.

Saunders unpicked this story, and others, in his book:

> Snowballs, a notorious brand of Ecstasy, consisted of very strong pure MDA which came from a government laboratory in Latvia . . . Latvia needed western currency and had the advantage of no drug laws, so they joined up with a German businessman to produce MDA for export as Ecstasy. This went well for a couple of years until a consignment of 10 million tablets was intercepted in Frankfurt airport, since when MDA has been rare.

Saunders' book was published a full year after the Snowballs first appeared. In the intervening period users had no idea that they were taking a drug other than MDMA, and even in the immediate years after publication the news was shared only among those who had read Saunders' book. Rumours often circulated when strong new batches of pills came on to the market that they were cut with ketamine or even heroin, the latter a particularly bizarre and unlikely choice considering its high price. Even the basic action of MDMA was widely unknown; the drug can be overwhelmingly sedative during onset if dosed incorrectly high, or if purity is greater than that to which users have been accustomed.

One commonplace belief among Ecstasy users at this time was that the drug caused the spinal fluid to drain away, leading to mass paranoia on comedowns when users were suffering from nothing more than a stiff back caused by hours of hectic dancing. The drug caused Parkinson's, it was rumoured, and left gaping holes in the brain: both untrue. Misinformation was rife, with Ecstasy users having, in the pre-web era, no simple way to research the drug or share any information they had. Such a situation would not and does not happen today. At the same time, pill manufacturers started to cash in by copying previously popular and clean logos and selling fake or tainted products, making the market even more dangerous.

Nevertheless, a new, hedonistic hegemony took root irresistibly in Britain, the US, Europe, Australia and Asia. *The Face* magazine in 1990 dedicated a twelve-page feature to clubbing in Europe, as the whole continent fell in thrall to the new groove.[17] From the shrink's sofa to the dancefloors of Dallas, New York, Detroit, Chicago, Ibiza, Manchester, Blackburn, Liverpool, Nottingham and London, and then the whole world, Ecstasy became the drug of choice for millions of people who had never got high in their lives before.

In comparison to binge-drinking and street-fighting, Ecstasy seemed to be an almost healthy and active lifestyle choice, with ravers turning away from alcohol as they sipped on energy drinks and dressed in leisurewear. The sensations those chemicals elicited in users, too, along with the glorious rush of unified, rhythmic euphoria, seemed less than harmless; they actually felt beneficial. Excessive use of the drug, as well as the adulterants that unscrupulous or unskilled chemists used to pack out the tablets, offered less of a tonic, and an extremely small percentage of users of the drug died from it each year. There were five deaths in the UK in 2011 from MDMA, according to the Office for National Statistics.[18]

But the drug has been used by many millions of people for decades now, and no long-term damage has been noted. Indeed, if a pure, dose-measured supply were freely available, many of the problems associated with excessive use would be wiped out instantly.

The market has behaved in ways predictable under classical economic theories. In 1988, when the number of consumers was relatively small, a tablet of Ecstasy cost fifteen to twenty pounds, and quality was generally high, with few 'brands' of tablets available. Chemical nostalgists will smile at the mention of Yellow Calis, New Yorkers, White Burgers, White Doves or Disco Biscuits (so named for their huge size and muddy, digestive-like colour). These drugs were expensive, but they worked: one or two pills kept users dancing for a whole night. Taking into account inflation over the last twenty years, the same pills today would cost around fifty pounds. But in the latter half of the last decade, around 2005, by which time the market had grown enormously, reasonably good-quality Ecstasy pills could be bought for as little as two pounds each, even in small quantities. The laws of supply and demand elegantly brought the price down, as quality dropped only slightly.

The arrival of Ecstasy in British drug culture has had a series of long-lasting and wide-ranging effects, but some of these are only really being felt today. Some revolutions occur in slow motion. In his social history of post-punk music, *Rip it Up and Start Again*, writer Simon Reynolds details the long-term impact of punk on the American and British musical scenes:

Revolutionary movements in pop culture have their widest impact after the 'moment' has allegedly passed, when ideas spread from the metropolitan bohemian elites that originally 'own' them and reach the suburbs and the regions. For instance, the counterculture and radical ideas of the sixties had far more currency in the mainstream during the first half of the seventies, when long hair and drug-taking became more common, when feminism filtered through to popular culture with 'independent women' movies and TV shows.[19]

The parallels with the way Ecstasy has infiltrated mass culture are striking. Initially the drug was the preserve of a well-connected metropolitan clique of pop stars, photographers and other creative people. After 1988, Ecstasy, an extraordinarily powerful drug, for all its hugged-up frivolity, colonized the leisure culture like a bacterial bloom in a petri dish. Ecstasy ushered in a completely new era of drug use in the UK, the US and Europe. Normal people with regular jobs were now routinely taking the most extraordinary chemical concoctions at weekends. More than ever before, the counterculture became the culture. In the UK an estimated 500,000 people use MDMA – when they can get it – every week. And when they can't, they'll just take something else. The aftershocks of the MDMA cultural invasion can now be witnessed in the contemporary rise of new and largely untested drugs such as those mentioned in the post by Clapham Boy in the introduction to this book.

Ecstasy was not the only thing to produce a dramatic cultural shift in the nineties and noughties. Just as Ecstasy use developed from the niche interest of an illicit underworld to become the dominant culture, so too did the internet. And the drug culture and the technology not only nodded across the dancefloor at each other, they shook hands and embraced.

# 3

## *The Birth of an Online Drugs Culture*

The first thing ever bought or sold on the internet was marijuana. The deal was done in 1971.

The history of the internet is bound up with the counter-culture, and the counterculture finds some of its richest expression in the use of psychoactive chemicals. Technological protocols and cultural pipedreams aligned and collided. The drug and music countercultures and the early technological innovators informed and inspired each other – and were often the very same people. The acronymic utopias enabled by internet technologies such as TCP/IP aren't so different from those offered by LSD: equality, connectedness, awareness of life as a sum greater than its parts.

In the early 1960s, American computer scientist Leonard Kleinrock of the Massachusetts Institute of Technology and Paul Baran of the Rand Corporation, and, later, Britain's Donald Davies, a physician at the UK's National Physical Library in Teddington, independently conceived of the same way to send data around a telephone network efficiently by splitting it into chunks and routing it through nodes around the network to later arrive, reassembled, in the right place.

These deliberate first steps towards cyberspace had a greater impact on the history of mankind than the simple stroll on a rock high above our heads two years later. This 'packet-switching' concept was to become the central structure in international telecommunications and, later, data networks.

Four months after the moon landing, on 29 October 1969, the Advanced Research Projects Agency Network (ARPANET), the world's first packet-switching data network, consisting of four computers in separate university sites, jumped into life. The first message ever sent was meant to say 'login', but the system crashed, and the first word ever sent from one computer to another was the accidentally portentous 'Lo'. As an opening line it's a little more truncated than Samuel Morse's famous dash-dot message, 'What hath man wrought', sent from the Supreme Court chamber in the American Capitol building in Washington DC to Baltimore's Mount Clare railway station in 1844, but the meaning was essentially – if unintentionally – exactly the same.

ARPANET is often described as the birth of the internet, and is equally often reported to have been designed to survive a thermonuclear strike, meaning that if one node or cell of the network were destroyed, the others would gather the digital slack and reroute the information around the surviving nodes.

However, the aim of ARPANET was not to preserve national security in the event of warfare, but to allow university researchers separated by geography to share information; the net's roots were indisputably collaborative and altruistic. Its technological cornerstone – the packet-switching network – underpinned all the later digital developments that would enable the reeling madness and quotidian mundanity that comprises a day online today – a day that includes buying groceries, paying bills, sharing photos and ideas, updating the world on your latest hairstyle choices, and, for many more people than is currently acknowledged, talking about and buying drugs.

Few involved in the early days of the internet could ever have imagined how central to billions of people's lives it was to become, but some of them dreamed of it. A year before the ARPANET came online, on 9 December 1968, Doug Engelbart, the ultimate unsung conceptual, philosophical and practical pioneer of modern computing, addressed a crowd of 1,000 programmers at Stanford Research Institute in Menlo Park, California. It was an event that was to become known as the Mother of All Demos, and during it Engelbart displayed publicly, in one gargantuan techno-splurge, many of the concepts of computing that are so ubiquitous today: the mouse ('I don't know why we call it a mouse. It started that way and we never changed it,' Engelbart said that day), video conferencing, hypertext, teleconferencing, word processing and collaborative real-time editing. It was the beginning of the modern age.[1]

Engelbart, in common with many intellectuals and technologists of the era, had attended LSD-assisted creativity sessions in the 1960s at the International Foundation for Advanced Study, a California psychedelic research group founded by a friend of Alexander Shulgin's, Mylon Stolaroff. The Shulgins wrote the preface to Stolaroff's book *Thanatos to Eros* (1994) detailing his experiences with LSD, MDMA, mescaline and a number of Shulgin's creations.[2]

Author Stewart Brand, who coined the phrase 'Information wants to be free' in 1984, was responsible for filming the Mother of All Demos, and that same year he launched the *Whole Earth Catalog*, the ad-free *samizdat* techno-hippy bible. Its esoteric and wide-ranging content, from poetry to construction plans for geodesic domes by physicist Buckminster Fuller, from car repair tips to trout-fishing guides and the fundamentals of yoga and the I-ching, was hacked together using Polaroid cameras, Letraset and the highest of low-tech. It now reads much like a printed blog; it was a paper website, in the words of blogger and author Kevin Kelly, that was sprinting before the web even took its first shaky steps.[3] Its statement of intent in its launch issue reads like a manifesto that has been realized by today's web users: 'A realm of intimate personal power is developing – the power of the individual to conduct his own education, find his own inspiration, shape his own environment, and share his adventure with whoever is interested. Tools that aid this process are sought and promoted by the *Whole Earth Catalog*.'

Brand, whose collaborations with Ken Kesey's Merry Pranksters would evolve into the Acid Tests, the 1960s proto-raves fuelled by LSD and documented by Tom Wolfe in *The Electric Kool-Aid Acid Test*, felt that information technology was the next stage in humans' evolutionary progress.

Info-anarchists and cyber-utopians not only laid the foundations for the internet, but would act as outriders for the free software movement. The net's founding mothers and fathers wanted to share their knowledge, and everyone else's knowledge, all at once, all the time, for free, with no centralized control system. Instead, they preferred – and created – a devolved, leaderless model of equalized authority. It was a computer in Engelbart's Augmentation Research Center at Stanford Research Institute that would make the second node of the ARPANET. His dream was a future where workers would sit at personal computers connecting and collaborating.

At the same time as many social hierarchies were being challenged, the technical architectures and hardware that would become the internet were taking shape. The links between the 1960s Californian freak scene and the pioneering days of early personal computing are chronicled in John Markoff's 2005 book *What the Dormouse Said: How the Sixties Counterculture Shaped the Personal Computer Industry* (even the book's title is taken from a hoary old Jefferson Airplane track). In it, Markoff revealed that the world's first online transaction was a drug deal: 'In 1971 or 1972, Stanford students using Arpanet accounts at SAIL engaged in a commercial transaction with their counterparts at MIT. Before Amazon, before eBay, the seminal act of ecommerce was a drug deal. The students used the network to quietly arrange the sale of an undetermined amount of marijuana.'[4]

In 1979, a worldwide discussion system called Usenet was

launched, and became the first step in the creation of the technical and cultural infrastructure that would personally connect humans across the planet, gathering enthusiasts and specialists into one digital conversational space.

Usenet was massively popular with about twenty million users in the early, pre-web 1980s and 1990s, and in many ways it was the first true example of social media, hosting thriving discussions on thousands of topics, known as newsgroups. It resembled a cross between an email and a web forum, similar in style and intent to such contemporary web communities as Mumsnet.

Usenet historians at Giganews.com note: 'Usenet was not only an important technical development; many social aspects of online communication were introduced, refined, and became de facto standards thanks to Usenet. Emoticons, flame wars, trolls, signatures, and even slang acronyms (BRB, LOL) found their first common usage on Usenet.'[5]

Users could not access Usenet's newsgroups unless they knew rudimentary coding skills. The system required knowledge of the Unix command line to set it up – there were no Windows-based computers at that time, so you couldn't simply type the name of the group you wanted to join or read, or click on an icon to access the software. Information technology and by extension, the internet, was for geeks, by geeks. And those geeks required bottomless oceans of patience, as the information trickled down the copper wires of telephone cables at a rate of bauds rather than megabytes.

As the service became more popular, Usenet groups became more chaotic and information harder to find and categorize. In 1987 a group of users known as the Backbone Cabal reorganized the service into hierarchies of interest, in a process known as the Great Renaming. These top-level hierarchies were: computers, news, scientific subjects, recreational activities, socializing and talk, with 'miscellaneous' covering the rest. In time the subsections in newsgroups covered every subject known to woman and man, and in an age before search engines, they were one of the best ways to find information online – or anywhere, since they were populated by the most extraordinarily helpful, altruistic and technologically adept users in the world.

Computer scientist Brian Reid, and John Gilmore (an early internet pioneer, civil libertarian, entrepreneur and techno-renaissance man whose work around cryptography, censorship and drug law reform make him an unsung early hero of the digital age – and who is also, judging by his love of tie-dye and several anecdotes, no stranger to a dose or three of psychedelics) felt the reorganization would limit freedom of speech. Gilmore was refused permission to create a group named rec.drugs, and later, talk.drugs. Reid was unhappy at the renaming of a food group he ran, and so together the two men decided to use technology to achieve their goals. They found a way to create a new, top-level hierarchy that that did not require the permission of the Backbone Cabal and that would be accessible to anyone with a modem and Unix experience. It was called the .alt hierarchy. Not only could users read .alt groups, they could create their own .alt sub-groups. No one could grant or refuse you permission. By creating a new Usenet hierarchy that ironically neither accepted nor required leaders, as other groups did, Gilmore and Reid fostered true freedom of speech in cyberspace. 'The net interprets censorship as damage and routes around it,' *Time* magazine quoted Gilmore as saying in 1993, predicting twenty years ago changes that would come to affect our legal system and our entire way of life.[6] The .alt hierarchy would be a free zone, and it was here that the earliest online drug culture formed.

The groups hosted within this network were some of the most popular digital watering holes for the outré and avant garde. Here, the sacred and the profane met: priests, poets and librarians communicated, perhaps for the first time, with perverts and potheads, and the early online drug scene began to coalesce around the sub-groups alt.drugs, alt.drugs.psychedelic and alt. drugs.chemistry. The author of the FAQ for those wanting to establish an .alt newsgroup wryly nodded at most people's mistaken assumptions about the new system: "'ALT' stands for "Anarchists, Lunatics, and Terrorists".[7] It was a joke, but the atmosphere in the .alt groups *was* ludic and countercultural to the point of cyber-anarchy.

One of the main attractions of newsgroups, and what made them so popular and functional, was that they not only enabled social interaction, but also allowed information to be collated in one place. As universities linked up to each of the nets, the main problem people had was how to actually find what they were looking for. Their imagination, and net use, was limited by the size of the data pool, and the lack of any clear directions. There were no indexes or catalogues, and connection speeds were treacle-slow. This was a period before file formats such as the jpeg were widely used, when ASCII was the global lingua franca and attaching large documents to emails was considered an inconsiderate use of global bandwidth.

The indexing systems that we take so much for granted – search engines – are central to all web users' daily experiences now. But in the early days of the web, finding information was a complex task carried out only by the skilled and the professional. Systems such as Gopher, ARCHIE, VERONICA and JUGHEAD were as unattractive and mystifying as their capitalized acronyms.

The domain name system, which uses words instead of numbers to request net pages from servers, was created in 1984. The curiously human Macintosh, by Apple Computer, was also launched that year, its use of icons and an onscreen cursor suddenly bridging the gap between person and machine, and bringing Engelbart's mouse to the masses. Revealing the Macintosh to the world for the first time in 1984, in a presentation that was to become the archetype for the company's hype-heavy launches, Apple boss Steve Jobs shocked the audience as he showed that the computer could speak, its rudimentary voice-emulation software ringing around the hall, sounding for all the world like a disembodied, time-travelling Stephen Hawking. For those who had seen Engelbart's demo, though, Jobs' entire presentational schtick looked more than a little familiar.

In the early 1990s, Stuart Brand set up The Well, a legendary bulletin board that was an early gathering point for intellectuals and cyberutopians. The Well, or Whole Earth 'Lectronic Link, was a virtual community that hosted conversations between some of the web's earliest champions including John Gilmore. It was also an important meeting point for fans of the Grateful Dead, confirming for ever the association between high-tech geekery and psychedelics that would result in the virtualization of the illegal drugs market in the twenty-first century.

Newsgroups disseminated the solid information on drugs that had been so lacking earlier. The Usenet newsgroup alt.drugs generated about 130 posts a day, its online FAQ said in 1995, and had about 120,000 daily readers. Alt.drugs. chemistry, a related group, began in 1994, and its sole topic was the covert manufacture of illicit drugs. It seemed unbelievable at the time that all across the world drugs policy was toughening in response to the new wave of designer drugs, such as crystal meth and Ecstasy, yet 120,000 inboxes each day received a series of innocent-looking text files that documented in meticulous detail how best to manufacture or take illegal compounds.

Where once drug manufacture was a completely hidden science, the preserve of motorcycle gangs, hippy renegades and organized crime syndicates, now this arcane information took its place comfortably among the reading matter of a technically literate avant garde. The net democratized criminality – or information that would enable criminality – on an unprecedented scale. Both the quantity of illicit information, and illegal acts inspired by that knowledge, were set to grow.

The FAQ to alt.drugs lined up the tattered and tie-dyed 1960s and 1970s drug myths – like the belief that smoking banana skins got you stoned, or that some LSD contained strychnine – and deconstructed each of them with logical prowess and, more importantly, trustworthy, hyperlinked sources. It also coined now-popular email slang such as IMHO (in my humble opinion) and WRT (with respect to). The community's view of drug laws as unwelcome intrusions into people's private lives has now become a much more commonplace belief.

As a virtual space where identity was shredded into binary code and reassembled as text on the screens of strangers separated by thousands of miles of space, but sharing closely aligned philosophies, the net was well suited to those interested in the psychedelic and psychoactive experiences. No laws seemed to apply there, and the early days of digital interconnection were characterized by behaviours and value systems that would have ended in a jail sentence had they been enacted in reality. To paraphrase Peter Steiner's famous *New Yorker* cartoon of 1993, if no one on the internet knew you were a dog, then equally no one on the net knew you took or synthesized designer drugs.

In 1996 an extraordinary series of threads started by the posters Eleusis and Zwitterion in the alt.drugs.chemistry newsgroup argued out in obsessive detail how best to manufacture MDMA. It was a twisted flame war, a soap opera for a voyeuristic digerati. But before he grappled with Zwitterion, Eleusis had first taken on the underground press classic *The Secrets of Methamphetamine Manufacture* by Uncle Fester, a pseudonym for the American clandestine chemist Steve Preisler, the father of modern meth-amphetamine manufacture.

Fester's most famous work was written after the electro-plating technician was jailed for three and a half years for possession of methamphetamine in 1984. The DEA said he was guilty of more than possession and pinned him with a synthesis charge after they produced evidence that he had been buying ephedrine tablets, a known precursor to the potent stimulant that has ravaged America's rural heartlands. While in jail, Preisler, nicknamed Uncle Fester by college friends after a TV character in 1960s comedy show *The Addams Family* who liked to cause explosions, decided to write the book as an act of defiance – and more importantly, to spread his ideas. 'My thought at the time was, "You don't like what I'm doing, huh? Well, how would you like 30,000 more just like me?" I know how to dig through the scientific literature, but I also know how to tell a story. There's nothing worse than a dull chemistry book, cause it'll make your teeth hurt,' he told Fox News in an interview in 2004.

If you skipped the rather intractable synthesis sections – and if you could overlook the fact that he was propagating information that could end in severe addiction and death – Preisler's book was a hilarious and extraordinarily individualistic declaration of the cherished American constitutional right of freedom of speech. American publisher Loompanic has had the book in print for decades now, and it is now in its eighth edition. Investigators claim it has been found in many drug manufacturing laboratories around the world.[8]

Eleusis felt that Uncle Fester's book was riddled with technical flaws, and so took it upon himself in 1996 on the alt.drugs. chemistry group to berate the author and destroy his work, chapter by chapter, line by line, with sources proving his case. Under the thread title: 'A Thorough Thrashing of Uncle Fester's *Secrets of Methamphetamine Manufacture* – No wonder the Feds don't give a shit!', he mocked and teased, scorned and shamed the father of modern methamphetamine manufacture.[9]

The boisterous Uncle Fester was goaded into defending himself against the allegations, and it made for great sport. It was a marvellous meeting of minds, but more than that it was the meeting of two worlds – the spluttering old guard overthrown by this vocal, technologically adept young chemist. Eleusis was just as defiant as Fester had been in print, but with a much larger audience than the old master. Furthermore, this new audience could replicate and distribute and share the information they received instantly and infinitely and at no cost. 'This book is an obvious attempt at profiteering,' concluded Eleusis on 3 March 1996:

> Perhaps Uncle Fester goes by this name in order to escape harassment by the Feds, but I'd say he would have more to worry about from disgruntled hack chemists that attempt to duplicate his 'work'. As well, it is quite clear that old Fester didn't do most of what he writes about, because essential details are lacking at every step of the way. This may be intentional omission to reduce liability, but if so, why bother? It is my opinion, then, that this book review is worth far more than the book itself. Enjoy!

An amused audience worldwide tuned in to this bizarre new entertainment channel. The conversation made little sense to many, but its impact was clear enough. Media was changing, as the means of production and, crucially, distribution were quietly seized. This would change the drug culture, in common with every other area of human experience.

Eleusis was a fascinating character to anyone interested in the drugs subculture and net culture at this time – his arguments with Zwitterion were technically impenetrable, but full of humour and clever literary and mythological references to ancient Greece and Homer. But Eleusis and Zwitterion were, it turned out, the same character. Floridian Jeffrey Jenkins was an English graduate who had turned his hand to basement chemistry in order to perfect the art of MDMA manufacture. He says he posted under two names to ensure a decent conversation, to increase the sum

knowledge of the technical processes and to avoid detection. It
may be that his frenetic creativity found expression in creating
believably distinct alter egos, or it may be that he was struggling
with the multiplicities of identity that self-representation in the
virtual space triggered in those pioneer days.

A text file posted to alt.drugs.chemistry after Jenkins' arrest
for manufacturing MDMA offers a fascinating insight into the
mindset of a typical, small-scale underground chemist, who is
obsessive, altruistic, individualistic and contrary, but passionate
about his work. There is something in both the MDMA experience and the net itself that fosters the innate human desire to
share the knowledge we have gained, to scatter the seeds of our
experience far and wide. 'Eleusis, for those of you unfamiliar . . .'
wrote the chemist after his capture and before his imprisonment,

   . . . was the name of an ancient Greek city where the
   Spring Mysteries were held: a city-wide festival where
   consumption of mind-altering substances was the central
   activity in a celebration of the return of Spring.
      Organic chemistry intrigued me. It tempted me with
   its secret language of symbols, its demand for (nearly)
   blind faith in unseen collisions. MDMA intrigued me as
   well, with its strangely universal experience, its ability to
   make even the hardest soul empathic. I had tried neither
   organic chemistry nor MDMA, so I decided to try both.
   In the Spring of 1994, appropriately enough, I began my
   chemical journey and by late winter I was already posting
   to a.d.c. [alt.drugs.chemistry] It took so much work to
   learn how to make MDMA that I decided I was going
   to share what I learned so that others would not have
   to repeat my labours. However, I had serious misgivings
   about sharing because my quest was one for knowledge
   and experience while, I knew, for most others it would be
   for purely economic reasons. You can see my struggling in

practically ever post I made, the schizophrenic vacillations in tone between erudite dissertation and egomaniacal evisceration. Though I knew my posts would be put to use by those less scrupulous, I posted nonetheless for the benefit of those who were.[10]

His experiments would ultimately cost Jenkins' family half their home in legal fees, and the chemist himself a long spell in jail. But in notes written after his arrest, he surprised many when he stated that the police had not questioned him about his activity on the newsgroup, and that his thousand-plus postings there had played no part in his arrest.

For those who might question the ethics of sharing information on drugs manufacturing online, the FAQ for the alt.drugs list – which had a casual approach to international drug laws – ended with a surprisingly idealistic and conventionally moral conclusion, written by Yogi Shan, another Eleusis pseudonym, some claim (the truth may never be known):

No matter how you rationalize it, there is no way to escape the cruel reality that drugs are about two things: money and power, amassed through the corrupt exploitation of human weakness. Sound public policy is built not through the cynical manipulations of politicians and two-dollar moralists, but through a careful balancing of harm minimization to the individual, _as well as_ society at large. Until society comes to grips with that, the non-medical use of drugs will remain an intractable scourge that distorts entire economies, corrupts our institutions to the core, and frays the social fabric. However, the base hypocrisy of society cannot and does not provide moral justification for the manufacture and distribution of illicit drugs for personal profit. Sorry.[11]

The author of the FAQ also lamented what he saw as an end to the halcyon days of Usenet as a resource for research and community building, as the service became populated by people with little experience:

> Usenet at its best is a network of some of the brightest minds in the civilized world, getting together to discuss whatever strikes their collective fancy. Professors and academics, engineers and scientists, polymaths, and intelligent people everywhere, getting together to kick ideas, information, and scurrilous personal attacks back and forth. A synthesis of great minds and intellects, altruistically donating their time and effort in glorious cosmic synergy. However, it's sad to say that, as more and more people go online, the Net is beginning to reflect the tawdry conglomeration that is society at large. One mammoth, lowest common denominator, vainglorious, pseudo-intellectual whore-house. To put it simply, Usenet may already have peaked.

He needn't have worried, though. The net drug scene was about to mutate once more, and technology was the driver.

During the Usenet era, the net itself had been changing – morphing into the world wide web, the global graphic interface to the new world of data invented and named by Tim Berners Lee at the European Organisation for Nuclear Research (CERN), in 1991. With great modesty and foresight, the physicist demonstrated and distributed, for free, a technology that would help his fellow particle physicists to share their findings into the fundamental nature of reality at CERN. Berners Lee's genius was to write HTML, or hypertext markup language, which allowed the linking of one document to any other that was hosted on the new networks. For the first few years, this new web would be popular only with expert users, but from 1993, with the launch of the Mosaic web browser, the technology caused a serious commercial and cultural buzz beyond academia.

During the alt.drugs.chemistry scene, one new drugs website called the Hive was also becoming a very busy online gathering point for people interested in the psychedelic experience, and in particular MDMA users and manufacturers. With its friendly buzzing bee logo, it was an early gathering point, from around 1997. Members, or bees, would talk about the synthetic routes to creating MDMA, or 'honey', as they would refer to the freebase oil the substance takes before it is salted into a solid crystalline form. At its peak, the Hive had over 6,000 collaborative members, and, as was so often the case in those days, framed its mission in terms of free access to information.

Instead of being hosted on Usenet, the site was hosted on the web in clear sight of authorities. It had threads – or discussion topics arranged into threaded debates – with hyperlinks, images and references inline in the text. It was a far more user-friendly, media-rich environment than earlier digital communities and users capitalized on the new possibilities immediately.

> The Hive is a discussion board with several moderated forums covering the whole area of the chemistry of mind-altering compounds . . . Many of these substances are subjected to strong legal restrictions in most countries. It is in your own responsibility to check your local laws and to apply for the proper permissions. Most if not all of the information discussed here can be found in public libraries, patent registers, or free internet sources. The Hive merely provides it as a compact collector's database.

Thus states its home page. Or rather, stated, since the site was taken down in 2004 after it became the subject of a ten-month *Dateline* investigation by American news channel NBC, which culminated in the unmasking of the site's owner. 'Strike' turned out to be the pseudonym for a chemical supply worker, Hobart Huson, who was subsequently imprisoned for eight years in 2003 for supplying drug labs with precursors, reagents and glassware. He was released in 2009.

The investigation was a toe-curling, voyeuristic affair which made much of young people's blurry sexual boundaries under the influence of the drug. The chemists were foolish enough to video themselves partying while high, and showing off new deliveries of glassware and lab equipment to the camera, much like teens today incriminating themselves on their Facebook pages.

In one scene of the documentary exposé, a journalist, agog at the fact that the constitutional guarantee of free speech also applies to those talking about drug synthesis online, has his lament echoed by a portly, moustachioed, cartoon-caricature DEA agent. The contrast between the young net evangelists who believed that information on any topic should be free, no matter what, and the law enforcement authorities could not have been starker.

Although the Hive was taken offline, an archivist-moderator at the site, Rhodium, had carefully saved thousands of posts that catalogued information about the manufacture of hundreds of drugs, and the archive was widely disseminated across the net. The information lay there quietly on hard drives and file lockers as a .torrent file, waiting for just the right moment, just the right social circumstances and chemical conditions, to reappear.

Elsewhere on the web, other activists gathered and offered information about banned drugs. In 1996 Nicholas Saunders, the author of E for Ecstasy, established the ecstasy.org website, placing the full text of his book online for free. 'Ecstasy.org aims to gather and make accessible objective, authoritative, and up-to-date information about the drug Ecstasy (principally MDMA),' said the site upon its launch, and it soon had three million hits a year from visitors looking for non-biased information. The site also contained pill-testing data, providing chemical analyses of tablets sent in by users. Grainy jpeg and .gif images of pills bought in the UK and beyond were compiled slowly line by line on screens, a major step forward in harm reduction. The data revealed that much of what was sold in Europe and the US as Ecstasy was inert, mislabelled or plain poisonous. It was an ingeniously innovative model that would be followed by many other websites soon after.

An American counterpart, Ecstasydata.org, was set up in 2001 and is still running today. It started publishing results online in July 2001, providing a valuable service to the drug-using public into what was being sold on the street. Information like this can save lives, since the majority of pills sold as Ecstasy on the street do not contain MDMA, and can sometimes contain deadly drugs that are similar in effect, such as PMA, a drug that has killed dozens of users over the years. Because of DEA restrictions, the site's administrator explains, most American labs are not allowed to test street drugs submitted anonymously, since possession of a controlled substance without valid prescription or license is a crime in the United States – that is, both the lab and the individual would be acting illegally. Most test labs in the US are limited to screening urine for employment or enforcement purposes and because of the threat of closure, refuse to accept a tablet for analysis.

Ecstasydata.org's lab has been given special permission by the DEA to receive submissions of street drugs from across the US and internationally. The site cannot reveal, as Dutch test centres can, the quantity (in milligrams) of a drug found in a sample, because the DEA has an unpublished rule that licensed labs are not allowed to provide quantitative data to the public, as they fear such a level of detail would provide 'quality control' information to dealers and users.

In an email, the manager of the service argues that this is irresponsible:

It is our opinion, based on long experience, that substantially more detailed information could be made available to the public and to poison control centres through this type of system without increasing risks to the public about encouraging illegal drug use. It is our view that the American government should not only allow, but subsidise street drug analysis to help promote awareness of contaminated and mislabelled drugs among users, parents, teachers, and children; facilitate long-term data collection about street drugs for future retrospective review; and provide a public, reviewable resource for medical professionals and poison control centers to help provide care for those who experience medical emergencies related to street drugs such as Ecstasy.

The site publicizes the results of tests on potentially deadly pills to users of the site, which now number 700,000 per year, and also carries out work offline at dance events. 'When we see particularly dangerous pressed tablets, we try to publicize the results to the online communities that might be impacted by the drug,' says the site's owner. 'Also, [sister organization] DanceSafe uses the results to carry out in-person harm reduction work by printing out or having a digital version of the results to show people at large dance or electronic music events.'

At the same time as Usenet groups were hosting conversations on the synthesis and use of drugs in the mid-to-late 1990s, many sites also sprang up on the newly created web selling plant-based drugs, and users of these plants, known as ethnobotanicals, took to the website Erowid.org to document their experiences. Many of them referred to these and other psychedelic drugs as entheogens, meaning substances used in shamanic or religious contexts, in an attempt to frame their drug use as a spiritual, rather than hedonistic quest.

These drugs had of course been popular in the pre-net days. The use of natural plant psychedelics grew in popularity after the psychedelic revolutions of the 1960s and 1970s, and was an extension of the hippy antediluvian pastoral dream. In the 1990s, lawyer and activist Richard Glen Boire produced *The Entheogen Law Reporter*, a photocopied, subscriber-only publication distributed from California that declared its intentions as follows:

> Since time immemorial, humans have used entheogenic substances as powerful tools for achieving spiritual insight and understanding. In the twentieth century, however, many of these most powerful of religious and epistemological tools were declared illegal in the United States, and their users decreed criminals. The shaman has been outlawed. It is the purpose of *The Entheogen Law Reporter* to provide the latest information and commentary on the intersection of entheogenic substances and the law.[12]

Twenty-two issues were printed between 1993 and 1999. They contained learned legal essays on the finer points of American drug law and policy, and laid the foundations for an online drug culture that was inquisitive, sure of itself and fully conscious of the choices it was making. The authors presented drug-taking as a spiritual pastime, and deliberately framed the war on drugs as a war on nature, since banning natural products such as magic mushrooms and peyote was both harder to enforce and more difficult to justify rationally. What's more, there was precedent for the legal use of natural psychedelics, for the American government graciously permitted, under certain restrictions, Native American churches the right to continue their millennia-long use of peyote in ritual settings.

Boire, who is now a lawyer working in complex drugs cases, told me by email in 2012 why he had produced the newsletter:

I was a young lawyer and was fascinated by entheogens. The law surrounding them was and is very convoluted, and at the time many people did not know what was or was not permitted. Accurate information about entheogens was hard to find, sometimes harder than finding the entheogens themselves. I also saw lots of parallels between banned books and banned substances – they both change how you think, yet banning books is considered old-school totalitarian, while banning substances is largely accepted. I wanted to investigate this.

With their love of free speech and work as information activists, it was a natural jump for readers of this and similar publications – such as *The Entheogen Review* – to look online for information about drugs.

Since its inception in 1995, Erowid.org – the name, meaning 'Earth Wisdom', hinting at the site's hippyish roots – has grown to become the world's most important repository of information about the 'complex relationship between humans and psychoactives', and the single most important resource on the web for information about drugs.

Its home page is herbal-heavy: poppies, cacti, Egyptian blue lotus leaves, kratom, a Thai leaf used as an opiate-replacement, mushrooms, mystical Mexican sages such as salvia divinorum. Erowid hasn't evolved much in graphic design terms since it started, but there is little need for it to do so. The form here is user-generated content by the gigabyte, supplied for free by generous souls.

From its inception the site gathered users' reports on psychoactive plants and also synthetic drugs, split into sub-categories in its 'Experience Vaults': General, First Times, Combinations, Difficult Experiences, Glowing Experiences, Bad Trips, Health Problems, Train Wrecks & Trip Disasters. These trip reports documented the effects of hundreds of drug and plant experiences, and their interactions, synergies and contraindications. It was like a shaman's hut with a modem. The site was the first port of call for anyone looking to document, share or make sense of their experiences – or to grandstand and stake out the new frontiers of consciousness, for it has to be said that this intensely geeky, mostly male subculture could at times be incredibly self-regarding and pretentious.

Natural drugs such as nutmeg, mescaline-containing cacti, mushrooms and vines containing psychoactive substances were among the first compounds to have their effects documented online at Erowid. That most of them were ineffective, or unpleasant, did not bother users or readers much, and the so-called ethnobotanical market still has a multimillion-dollar turnover even today, since most of these plants and seeds remain legal.

Some reports hosted by the site detail acts that make the mind reel at their folly, while others bore to tears with their meticulous data-harvest. The reports by users who have eaten the oddly alluring flowers of *Datura stramonium*, though, are uniformly astounding. Why anyone would ever willingly take datura, which grows wild across the world and has been used for centuries in shamanistic contexts, is a complete mystery, but perhaps reveals the reckless lengths to which some people will go to experience a different state of consciousness. In the UK datura is known as thorn apple; in the US, limson weed. A member of the *Solanaceae* family, with spiky horse-chestnut-style seed pods and fluted, trumpet-shaped flowers, it is a fearful-looking plant. And nature hereby warns us that it is indeed poisonous: its seeds and petals contain the tropane alkaloids atropine, scopolamine and hyoscyamine, and render the user – or abuser – delirious for days. That's no hyperbole; scopolamine-containing plants are not hallucinogens; these-tropane-alkaloid-rich plants are classed as deliriants and send users into psychotically deranged states where their memories are obliterated, to be replaced with the darkest possible imaginings. Known in folklore to herbalists skilled in their preparation for use in treating ailments such as asthma and used worldwide in sacramental settings, in the digital age their consumption has been documented by the brave and foolhardy at Erowid.

The trip reports' titles alone are enough to put you off: 'A brush with death and total confusion'; 'Eating bugs while my friends convulsed'; 'I lost my pets and almost burned the house down'; 'A tale of nudity, arrest and insanity'. Most reports involve a trip to a psychiatric ward or emergency department after in-depth conversations with non-present friends, and a curiously universal endless search for imaginary dropped cigarettes.

Other new natural, powerful, plant-based drugs that became popular at this time included *Salvia divinorum*, a bizarre member of the sage family that catapults users into phantasmagorical and often unpleasant trips (quite often, they just fall over and hit their heads), morning glory seeds, Hawaiian Baby Woodrose seeds – all were revealed as natural and legal hallucinogens, and Erowid was instrumental in the widespread dissemination of this information, previously locked in books and journals. Vines containing DMT, as taken by Burroughs in the 1950s, were discussed and kits using them and other herbs to create ayahuasca, a potent DMT-containing jungle brew used by shamans in the rainforests of the Amazon, went on sale in the US and continental Europe, legally.

There is even a report at Erowid for sapo, a toad venom that is applied onto self-administered burns to the skin. Apparently it makes you feel as if you're dying for fifteen minutes, then users reportedly become stimulated and energized for a few days after that. Traditionally, it is used by hunters in indigenous communities in the Latin American rainforests, giving them a high resistance to fatigue and immensely sharpened senses, including, allegedly, the ability to see much further and to hear the footfall of prey from miles away. Its active ingredients are the peptides phyllocaerulein, phyllomedusin, phyllokinin, demorphins and deltorphins. Some of these frog venoms have been used as performance-enhancing drugs in racehorses.

Erowid's co-owner, known as Fire, told me she set the site up as 'a bit of an accident':

> In 1994, Earth [her partner] and I moved to San Francisco after graduating from college and were looking for interesting new jobs. I decided that web design sounded like an interesting and booming business, so I sat down to learn HTML and web page design. We both had an existing interest in psychoactives. We had made some attempts at scholarly research while in college, as well as joining a couple of email lists where related topics were discussed.
>
> While learning web design, our relatively meagre archives of information became an obvious source of data to practice making web pages out of. We put a few articles and pieces of data on a web page and then someone would ask a question in a discussion group that was answered on one of those pages. We'd point them to the page. A few more pages would go up and the URLs would get passed around. And it really was kind of just like that that Erowid was born. Then it snowballed.

To those who have criticized the site for providing information to drug users, Fire offers a calm rejection:

> It became obvious and is still clear that humans are not going to go back to a time where interested people don't have access to information about psychoactive drugs and technologies. The question must then shift to how to head from a dark age where prohibitionist policies intentionally tried to suppress information and pollute facts with political messages, into an age where we can

collectively be building a reliable wisdom base from which parents, teachers, and people of all ages can make informed decisions.

Daily, the site now gets 90,000 unique visits, and serves 4.1 million files. It contains over 60,000 public documents detailing case law and precedent in complex legal cases, and thousands of first-hand reports of psychoactive drug experiences. 'All reports go through a rigorous review process,' says Fire. 'So far we have published 22,000, rejected another 22,000, have 36,000 rated and ready for review, and another 13,000 yet to be looked at.' She adds that she receives at least a couple of messages a week from people who explicitly say that information they found at Erowid has saved their lives. Even today, the site is an invaluable resource for people taking new drugs that they have sourced from the internet, and its contribution to harm reduction is inestimable. It is also a valuable first reference point for parents, teachers and poison control toxicologists.

Just as *TIHKAL* and *PIHKAL* became required reading for many in the counterculture in the 1990s, Erowid became a resource for any early web user interested in drugs. In 1996, with the Shulgins' permission, the second half of *PIHKAL* was published online by Lamont Granquist, an early net advocate, who had created the Hyperreal Drug Archives, an early collection of files and information about psychoactive drugs. In 1999, Erowid moved on to the Hyperreal server and incorporated the archives there into Erowid, including *PIHKAL* and *TIHKAL*. While this was no real surprise to any informed observer, it still felt like a revolutionary act in an information war, which, in some respects, the war on drugs had become. To see Shulgin's complex psychedelic drug recipes published to the entire world, for free, was extraordinary. Now, not only could you read the information Shulgin had preserved so presciently, you could forward it to anyone with a few keystrokes.

Shulgin himself joined the online party in 2001, when his Ask Dr Shulgin site launched at American campaign group Cognitive Liberty's site. The Doctor took questions from members of the public in a weekly email Q&A and seemed to relish the contact with this new audience. The site ran for around three years and placed the counterculture's most articulate and learned chemist-hero at the heart of the online drug debate, with discussions focusing not just on complex technical matters, but also on the ethical and moral dimensions of the war on drugs.

Shulgin's influence and experience bridges the gaps between the early 1950s intellectual explorers and psychiatric treatment pioneers, the 1960s hippy counterculture, 1970s and 1980s underground psychiatry, the 1980s explosion of Ecstasy as a recreational drug, the early internet drug scene of the 1990s and early 2000s – right into the chaotic twenty-first-century situation.

What was to complicate the picture was a development that, on reflection, was entirely predictable. As the twentieth century ended, the web wasn't just a place where you could talk about drugs – it was about to become a place where you could buy them.

# 4

## The Rise and Fall of the Research Chemical Scene

Soon after *PIHKAL* appeared on Erowid, in around 1999, the ring-substituted phenethylamine and tryptamine analogues that Shulgin had made and tested on himself and his friends started to appear for sale on rudimentary websites in the US. Users online, where there was a rising wave of chatter about their effects, referred to the new drugs Shulgin had invented as 'research chemicals'. Research chemicals are nothing more than designer drugs – but drugs that until recently very few people had ever taken. They are broadly either hallucinogens, or empathogens (drugs which bring emotional insight), or stimulants; there are hundreds of them, and they produce almost as many and varying effects as grapes produce wine or milk produces cheese. To fully describe the subjective effects of each of them in turn would fill several volumes.

At this early stage these drugs were used by a few thousand self-defined 'psychonauts', or explorers of inner space, who researched their effects by browsing scientific literature and discussing them online. The internet facilitated the supply and distribution of these new drugs, whose names are a baffling alphabet soup of numbers and letters, but most critically it helped people find out which of them were fun, terrifying, transcendentally visionary, a waste of time or money, or fatal.

This was as much an information revolution as a chemical uprising.

Consider the dilemma of a research chemical user who could find no accurate information on dosage or the interaction between the new chemicals she has found or manufactured or had made. The safest and most rational solution would be to ask people who had done it before, and with the web that kind of communication became not only possible, but simple. Research chemical users took to Erowid in their droves, filing thousands of reports about the new drugs. Some were lengthy, Shulgin-inspired accounts, detailing their doses and their 'set and setting' – that is, describing the users' states of mind and their physical environments before they took the drugs, since both can impact hugely upon the psychedelic experience. From these inconsequential and underground beginnings, whose effect was felt solely by a tiny minority of reckless or fearless explorers, the virtualization of a part of the international drugs market began.

Research chemicals in 2000 were made in clandestine laboratories on a very small scale in the US, sometimes in Eastern Europe, and, more commonly, in China. The situation is broadly similar today, with China the world leader in their production. At the turn of the century, most of these drugs were legal to produce and export almost all over the world, as although many of them had exactly the same kinds of effects as banned substances such as LSD and certain amphetamines, they were so novel that they did not feature in international drug legislation such as the Single Convention on Narcotic Drugs, 1961, or the UN Convention on Psychotropic Substances, 1971, and nor did they feature in many countries' national drug laws. They were not specifically named in the American drug schedules, but they could arguably have fallen foul of the American Analog Act of 1986, of which more shortly.

British law in 1999, though, was rather more advanced than that of many other countries in chemical terms, and certainly far tighter than that of America. Whereas the US attempted to control the appearance of new drugs on the basis of their activity and chemical similarity to banned substances, the UK had crafted tightly written laws specifying exact molecular structures and ring substitutions that would make little to no sense to anyone but expert chemists. And those laws were about to get even tighter.

Swedish naturalist Carl Linnaeus, who was born in Råshult, Småland, in 1707, is the father of modern plant and animal classification. On an expedition to Lapland in 1732, Linnaeus travelled 4,600 miles across Scandinavia and then, on foot, across the Arctic Ocean, discovering 100 botanical species. In 1788, the Linnaean Society of London, the oldest biological society in the world, was formed. Its HQ, in Burlington House, Piccadilly, is home to one of the world's most extraordinary animal and plant collections, and its walls are lined with books and prints; it is a repository of knowledge and selfless exploration and investigation. On 12 February 1999, the London Toxicology Group (LTG), specialists who study the effects of poisons and drugs on the human body, met there to discuss research that would help ban many discoveries made by Alexander Shulgin, whose interior expeditions into unknown realms were now considered a danger to society.

In 1998, there had been a number of acute poisonings and fatalities at raves around the UK, due to a potent amphetamine derivative, 4-MTA. The drugs were being sold under false pretences by dishonest dealers as super-strength Ecstasy pills, and they were known as 'flatliners' because their dramatic and unpleasant effects most closely represented a coma – users would simply pass out after taking them, lying inert in corners of nightclubs or raves, a disturbing exercise in deliberate narcotic nihilism. The LTG discussed a number of cases from a recent rave in Shepton Mallet, called Dreamscape. The deaths sounded uniformly gruesome, with one man, twenty-one-year-old psychology student Rene Saunders, dying by the roadside alone, writhing in agony.

4-MTA was actually invented by another American academic, a chemist named Professor David E. Nichols, in his search for medicinal, non-toxic serotonin-releasing agents for use in therapy and as anti-depressants. Nichols is both an experimental medicinal chemist and a friend of Shulgin's and is, in many ways, Shulgin's heir. The foreword to *PIHKAL*, which Nichols wrote, ends with the line: 'Some day in the future, when it may again be acceptable to use chemical tools to explore the mind, this book will be a treasure house, a sort of sorcerer's book of spells, to delight and to enchant the psychiatrist/shaman of tomorrow.'[1] He has since claimed that line was hyperbole, praise for a much-loved friend's work, distancing himself slightly from the maverick's methods.

While the wild-haired Shulgin laboured in his squirrel-infested shack, the only rodents anywhere near Dr Dave, as he is known to many people, were test subjects in cages a few blocks away from his office at the campus on Purdue University, Indiana, where he occupies the Robert C. and Charlotte P. Anderson Distinguished Chair in Pharmacology.

His work, like Shulgin's, has been adopted by drug users and dealers and synthesized worldwide for recreational use. Nichols, though, is the straight man to Shulgin's outlaw; and while Shulgin might be faster on the draw, Nichols might just be the sharper shooter. Where Shulgin confronted the law, questioned it, and ultimately encouraged millions of people to ignore it, Nichols works from within the established medical system and does not proselytize as Shulgin did for people's right to use drugs. His aims are in some ways similar to Shulgin's, but his methods are altogether different.

With his full but neat beard and easy demeanour, this learned chemist is also a keen gardener and plays a mean blues harp.

He admits he moves in circles that would like to see the legalization of psychedelics for psychotherapeutic purposes, but his approach to the disciplines in which he has specialized for decades – medicinal and bio-organic chemistry, molecular pharmacology and toxicology – is meticulously and conventionally rigorous.

Like all academic chemists, after his work on 4-MTA had been peer-reviewed Nichols published it, on this occasion in the *European Journal of Pharmacology* in 1992.[2] Papers such as these are available to anybody, either archived in public libraries or nowadays, in many cases, online. In the late 1990s 4-MTA was found on the streets of the UK and in shops in the Netherlands. When it was taken recreationally its effects were slow to build, and so users simply took more pills, chasing the effects. Most collapsed, and some died, as they had at the Shepton Mallet rave. It was one of the first instances of the widespread use of research chemicals, though this drug, in particular, was not mainly sourced online, and was sold under false pretences. Five people died from use of the drug in the space of about a year.

4-MTA was initially sold in the 'smartshops' in Amsterdam, where psychoactive seeds, herbs and smart drugs, such as nootropics – supplements and non-licensed medicines that improve cognition and memory – were sold, legally. These shops also sold magic mushrooms, recovery kits that claimed to ease drug comedowns, Ecstasy-testing kits and other high-tech drug paraphernalia.

The smartshops stopped selling 4-MTA in around 1998, when it was found to be dangerous. Remaining stock was then smuggled to the UK, where a seizure of 25,000 pills, each containing 100 mg of the compound, was found in 1998. Many more got through, with fatal consequences.

Nichols explained on Erowid why he had originally made the compound:

We had been looking for drugs that cause the release of neuronal serotonin, with the expectation that they might have therapeutic value similar to the SSRIs [antidepressants similar to Prozac]. I am sad to see it being used on the streets. I can't imagine what pleasure it might produce in users, because our tests with similar compounds in rats showed the substances to have aversive or unpleasant effects.[3]

A user of the drug confirmed that in an Erowid report:

I was screaming inside my head but couldn't talk properly. I ran desperately around to try and get the experience to end, my mind was ruined. I knew I had gone too far taking this drug . . . amphetamines, psychedelics, even ketamine I can handle, but this was beyond recognition, beyond comprehension . . . I could hear voices telling me I was going to die and that I should probably end it all. I was sweating like a PIG. I felt like I couldn't breathe, everyone else was breathing air. But air wasn't right for me, I needed another gas. God, I felt lost. God was telling me I was.[4]

After the gathered scientists of the LTG had discussed 4–MTA, it was no surprise that, in common with their European counterparts, they recommended that it be controlled.

They then turned to the next item on their agenda: the impacts of a new and emerging designer drug craze that some of their members had identified, in which potent new psychedelic chemicals made by Alexander Shulgin were also being sold on the web. Minutes of the meeting note:

Dr Les King of the Forensic Science Service outlined the Misuse of Drugs Act and its application primarily to the phenethylamine group of compounds. All Class A phenethylamines are ring-substituted and are hallucinogenic. These are listed both specifically and generally under the Act. Other phenethylamines are found in Class B (amphetamine and methylamphetamine) and Class C (benzphetamine). The book by Alex Shulgin *PIHKAL (Phenethylamines I Have Known and Loved)* contains 170 ring-substituted compounds and these are covered under the Act as Class A drugs. Around 34 compounds listed in the book, together with 4-methylthio-amphetamine (4-MTA), were found to have escaped the general classification [in 1971 and 1977], but these are now to be covered by legislation under Statutory Instrument 1999.[5]

The earlier ban of phenethylamine and tryptamine compounds in 1977 – when the Misuse of Drugs Act had first been amended to outlaw many ring-substituted amphetamines following the discovery of the psychedelic drug bromo-STP in the Midlands – had missed dozens of Shulgin's compounds, mainly because his chemical ingenuity was more advanced than that of British lawmakers or government advisors.

But now Shulgin's books neatly did legislators' and advisors' work for them, by collating in one place with great accuracy all the remaining possible variations on the basic phenethylamine structure he had discovered until that point. The Advisory Council on Misuse of Drugs (ACMD), the independent body that advises the British government on drug-related issues, would draw on this discussion by the LTG, and its advice to government was to blanket-ban the rest of *PIHKAL* and *TIHKAL*.

In the event, thirty-six individually named drugs were added to the Misuse of Drugs Act 1971, and the rest of Shulgin's published work was outlawed in the UK, effective 1 February

2002. The drugs were categorized as Class A, alongside heroin, cocaine and other addictive substances, even though they are not, arguably, as harmful, and certainly in no way physically addictive.

However, the new law inadvertently left the door open for any other drug that did not match these descriptions. Like water, drug designers would soon find their way around these obstacles. Indeed, the obstacles, if viewed from a certain perspective, conveniently outlined with great clinical, pharmacological and legal exactitude exactly what was legally permissible.

On the other side of the Atlantic the loopholes were not only wider, they were created by an opaque law. The USA did not have a catch-all ban on the work of Shulgin, and instead relied on the 1986 Controlled Substance Analog Enforcement Act, which attempted to thwart the work of criminal chemists by banning drugs that were 'similar' to banned substances. 'A controlled substance analog shall, to the extent intended for human consumption, be treated . . . as a controlled substance in schedule I,' the Act declared, defining an analog (analogue in the UK) as a substance that is chemically substantially similar to the banned drug it is based on, that has a stimulant, depressant, or hallucinogenic effect similar to the parent molecule, or is presented as such.

However, this blunt phrasing ignores the complexity of both organic chemistry and English grammar. To be illegal, need a new compound satisfy one, two, or all three of the requirements? The matter hinges on that clumsy, and innocuous 'or' at the end of the second clause, said Shulgin in *PIHKAL*, who added that the American law is written so unclearly as to appear to be a deliberate act of obfuscation, enabling legislators to jail people at will. In an area as complex as organic chemistry, it was legally and socially myopic.

Long before the new drugs market burst out online in 2000, Damon S. Forbes of Colorado was brought before the courts for purchasing alpha–ethyl tryptamine (A–ET), a medicine in the tryptamine family, from a chemical supplier online. A–ET works as an antidepressant if you take a low dose; ramp it up and you'll find yourself hallucinating. On 20 November 1992 Judge Lewis T. Babcock had the unenviable task of deciding whether A–ET was an analogue of DMT, under the Act. 'Defendants contend that this section requires a twopronged definition. The first prong requires a substantially similar chemical structure. The second prong requires either a substantially similar effect on the human nervous system or the *intent* [my emphasis] to have such an effect. The government argues that a substance may be an analogue if it satisfies any of the three clauses. I agree with defendants,' he said. 'I hold that the definition of controlled substance analogue as applied to A–ET under the unique facts here is unconstitutionally vague. Without doubt, it provides neither fair warning nor effective safeguards against arbitrary enforcement,' he ruled.[6] The case was dismissed.

Charges based on chemical structure and effect could be successfully challenged if you had good enough lawyers. Proof of intent to supply for use as a drug could be argued away with a label stating 'Not for human consumption' – a move of transparent sophistry that would partially inspire a later wave of net-based designer drug vendors to label their goods as 'plant food' or 'bath salts', winking at prospective buyers while dodging drug, food and medicine laws.

The inadequate patchwork of international legislation around synthetic drugs was soon to be weakened still further by the connective power of the web. And no matter what the laws did or did not say, there was a sense in the US – and beyond – in 2000 that online, drug laws didn't really apply, especially when the chemicals people were buying did not feature on any banned list of substances anywhere in the world. After years of synthesis discussion on Usenet and at the Hive, by 2000, as web use grew and after the publication of *PIHKAL* online, the research chemical business was starting to take off. You could buy many of Shulgin's psychedelics and have them delivered anywhere in the world in just a few days. The lid was lifted on the psychedelic treasure chest of *PIHKAL* and *TIHKAL* and hallucinogens and empathogens such as 2C-I, 2C-E, 2C-T-7, 5-MeO-DMT, 2C-D, 2C-T-2, and AMT became available to anyone with a credit card. Research chemicals were sometimes sold in very small doses and so large stocks were not necessary to maintain decent inventory: 1,000 mg, or one gram of 2C-E could quite easily be sold as 100 separate and powerful 10 mg doses.

An early forerunner in the American research chemicals scene in the late 1990s and early 2000s was JLF Poisonous Non-Consumables, which had a bold *Amanita muscaria* mushroom on its home page, but links on pages beyond that brought up dozens of different compounds skirting legality, including 2C-T-7, a drug Shulgin noted for its colourful hallucinations, and which his research group responded to almost universally positively. One of them wrote:

**PIHKAL #43 2C-T-7**
(with 20 mg) I lay down with music, and become engrossed with being as still as possible. I feel that if I can be totally, completely still, I will hear the inner voice of the universe. As I do this, the music becomes incredibly beautiful. I see the extraordinary importance of simply listening, listening to everything, to people and to nature, with wide-open receptivity. Something very, very special happens at the still point, so I keep working on it. When I become totally still, a huge burst of energy is released. And it explodes so that it takes enormous effort to quiet it all down in order to be still again. Great fun.[7]

These drugs had seldom been tasted by humans (except by friends of Shulgin), so Erowid became an essential source of information on both their effects and the best means of ingestion – whether they should be smoked, snorted or injected, or taken as an enema (a popular method known as 'plugging', favoured both by those looking for stronger effects and by the thrifty, for doses are lower if absorbed by the membranes of the anus).

The sense of excitement and community was palpable during those years from 2000–4. New reports and new compounds were emerging with dizzying speed. Some of these drugs were being sold not only online but also in backstreet headshops, techno-hippy neon-lit caves, their hugger-mugger shelves also filled with the new nootropics such as piracetam, Chinese erectile dysfunction analogues and Oaxacan dreamherbs like *Calea zacatechichi*.

The scene quickly became global. Japan was an early adopter and headshops there, after years of stocking ineffective legal highs, started selling compounds that had previously only been seen in Shulgin's shack. As in today's anarchic scene, the new drugs were sometimes sold with labels that hid their true contents. 2C-T-7 was sold online and in the streets of Roppongi, Tokyo, in 18 mg vials under the brand name Blue Mystic Powder (its true formulation wouldn't emerge for years). People who bought it knew full well they were buying a drug, and shop-owners would advise, quietly, how it was supposed to be taken and what effects might be expected.

In an interview published in September 2012 in *VICE*, a designer drug manufacturer told journalist Hamilton Morris about the roots of that drug's appearance:

> Around 1998 there was a group of us that were trying to work on some of Shulgin's thio-compounds, the 2C-Ts. They were a lot more difficult than the standard phenethylamines and we just couldn't do it effectively. So eventually a private group of chemists and investors pooled their resources and commissioned a laboratory

in Poland to produce a kilogram of 2C-T-7. It was ridiculously expensive, and the entire process felt like a really extreme measure. To the best of my knowledge, that group effort was the first instance of custom syntheses of a gray-market drug by the end users. Less than two years later, the chemical took off and was introduced as Blue Mystic in the Netherlands, and then as a pure chemical [online] in the States. 2C-T-7 was one of the first 'research chemicals' in the modern designer-drug sense, and I think some of its initial popularity came from the fact that it had been totally unavailable due to the difficulty of producing it in a clandestine lab.[8]

The Netherlands was a European hotbed of the early research chemical scene, which existed as much on the streets there as it did on the web. A drug marketed as Explosion was sold diluted in bottles of so-called 'room deodorizer' in Holland. Their contents were later identified as a simple solution of BK-MDMA, or methylone – a sort of MDMA-lite that was created by Shulgin in the years after his books first appeared in print. (Nobody knew at the time, but it would have been legal to sell methylone in the UK, since its specific structural modification on the phenethylamine skeleton had been missed by both the 1977 and 2002 amendments.)

Methylone is similar to MDMA, but its effects are less dramatic, less profound. Some users say it is a more 'honest' version of MDMA, which they feel can engender what seems in retrospect a phony intimacy. For every user, there is a different experience, though, and because of this many posts on forums where these drugs are discussed ended with the disclaimer acronym 'YMMV' – Your Mileage May Vary. Psychoactive drugs are subtler than alcohol, and the state of mind of users and the environment in which they consume them can change the effects of the drug enormously.

The number of drugs on the market and their availability were expanding every few months in the late 1990s and the start of the 2000s. Some drugs and the groups using them were truly bizarre: one compound, DiPT, was said to make music sound as if it had dropped an octave, while another powerful tryptamine, DPT, was believed by members of The Temple of the True Inner Light, an esoteric religious movement in Manhattan, to be the literal flesh of God.

Online vendors started to commission custom syntheses in laboratories and fine chemical companies in China started to take over most of the manufacturing. The names of the Chinese companies began to be jealously guarded – as well they might be, for this was slowly becoming a multimillion-dollar business. Chemists were winning a game the authorities did not even know they were playing, and people were taking, making, buying and selling all manner of novel psychoactive compounds and talking about them openly online. And they appeared to be evading legal repercussions for the first time, re-routing around the laws stemming from Nixon-era diktats and trite Reaganite homilies as lithely and blithely as data squirms past censors.

The American war on drugs had created not only a market, but a motivation for both the sale and consumption of these drugs. If a drug user wanted to experience something akin to LSD without being jailed they could simply buy these neo-legal alternatives. Vendors believed, wrongly as it turned out, that the pseudo-scientific nomenclature of research chemicals and the enclosed instructions not to eat them would save them from the attentions of a 4 a.m. battering ram from the DEA. They believed that if it came to a court case, they could claim they sold the drugs as curios, collectors' items for chemistry geeks. JLF Poisonous Non-Consumables had a fairly comprehensive disclaimer that it thought would protect it from the authorities. Its mockery may have cost it dear:

Do not take orally (into your mouth) as a food, a beverage, a chew, a toothpick, a nutritional supplement, a medicine, a recreational drug or an agent of suicide. Do not inject, inhale, snuff, snort, smoke or slam. Do not stick, put, insert or throw into your or another person's mouth, nose, ear, eye, anus, urethra, vagina or any other orifice or port-of-entry that may exist on your or another person's body. Do not allow any carbon-based product to become moist, then allow it to decompose with a pathogenic micro-organism, then allow the foul-black-rot to come in contact with your body, (especially mucous membranes) or insert into the orifices previously mentioned, thereby causing an infectious disease. Do not do that. Also, do not do this: Do not deploy any of JLF's products as weapons of war or tools for violence such as dangerous high-speed projectiles aimed at people or property. Do not use for tinder to start a fire to commit arson or to burn yourself or another or any public or private property. Do not leave lying on the floor to trip over or slip on to incur personal injury.[9]

Within that rather too-pleased-with-itself disclaimer there was an interesting reframing of drug use as a set of choices for which each user is responsible, with the individual the sole arbiter of right and wrong. The subtext was clear: if you know what you're doing, if you have done your research, read the books and the sites and the trip reports, then why should the nation's drug laws apply to you? It, and more generally the entire research chemical scene, was startlingly individualistic and felt like a digital-age updating of 1960s philosophies.

But some did not use the new drugs correctly, and overdosed and died; the margins of error and safe usage were narrow. Several people died after dosing on 2C-T-7 incorrectly – they snorted it, which intensifies its effects since the drug is not metabolized first by the digestive system. These deaths led to the very first exposé of the research chemical scene, in *Rolling Stone* magazine, in 1999. In October 2000, a twenty-year-old Oklahoma man, Jacob Daniel Duroy, snorted about 35 mg of 2C-T-7. Within moments he was vomiting and yelling about being attacked by evil spirits. An hour and a half later, on the way to the hospital, he died of a cardiac arrest.

In April 2001, seventeen-year-old Joshua Robbins from Cordova, Tennessee, did not heed the most important of warnings repeated endlessly online: he too snorted an unmeasured quantity of 2C-T-7. He died in pain, screaming, *Rolling Stone* magazine reported, 'This is stupid, I don't want to die.' The same month an unnamed man died in the Seattle area after combining an unknown quantity of 2C-T-7 with 200 mg of MDMA.[10]

It was alleged that one of these victims died from a compound sourced from JLF Poisonous Non-Consumables. The site was shut down by the DEA in 2001. The owner, Mark Niemoeller, received a sentence of three years of supervised probation, twelve months of home confinement with electronic monitoring, a fine of US$12,100, and the forfeiture of US$200,000 and a vehicle. But it was a warning shot that few in the industry heeded.

In 2003, the DEA's Operation Pipe Dream arrested fifty individuals who sold paraphernalia such as pipes, bongs, rolling papers and other such tat online. DEA acting administrator John B. Brown III told journalists: 'One important facet of this case is the use of the Internet by drug paraphernalia marketers. There's no easier way to reach young people – and to get around their parents – than through the Internet. It takes a lot of hard work to get at these sites. But we can assure worried parents that today there are 11 dot.coms that are dot.gone.'[11]

Brown's vanity in believing that he had controlled the situation seems even more pronounced when viewed from the perspective of 2013.

The new drugs scene kept rolling for a few more years, but for every high, there's a low, and the ultimate comedown hit hard when it came. It is possible that the 2003 launch of the Research Chemical Mailing List (RCML), which aimed to facilitate the purchase of these new drugs by recommending companies not listed on the .alt forums, played a part in the DEA's discovery of this hidden scene. The RCML collated and aggregated information from trusted contributors and other sources and attempted to regulate what was, essentially, an illegal industry. 'This list is an attempt to bring a comprehensive, up-to-date listing of all research chemical companies that do not require DEA licensing from their consumers,' the admin announced in 2003. It went on, 'Because of the publicity research chemicals have gained recently, it is no longer safe to publish the names of these companies in public forums like alt.drugs.psychedelics. There are also new companies springing up all the time. Some of these will undoubtedly be fraudulent. That is why we are promoting a private dialogue between list members for the discussion of these companies, the grade of chemicals they sell, and their prices.'[12]

The RCML administrators were respected and despised in equal measure for the move, since it handed valued information out to anyone who subscribed. The research chemical market had, by 2003, grown worldwide to a multimillion-dollar business and presented lawmakers with a unique set of challenges. Users knew more than the police about the laws, and about the drugs they were using. The drugs themselves were so small they could be sent anywhere in the world. But money trails and paper trails were now added to data trails, and it was inevitable that the hammer would fall, and the party would end. On 21 July 2004, the authorities swooped. Operation Web Tryp was a DEA clampdown that resulted in the arrests of ten vendors and the closure of five research chemical vendor sites: Pondman.nu, American Chemical Supply, Omega Fine Chemicals, Rac Research and Duncan Lab Products.

The DEA announced its success – the war was over, mission accomplished. DEA Administrator Karen P. Tandy said:

> Operation Web Tryp investigated Internet websites distributing highly dangerous designer drug analogues under the guise of "research chemicals" primarily shipped to the US from China and India. These websites are known to have thousands of customers worldwide. One website operator is known to conduct estimated sales of US$20,000 per week, while another is known to have been in business for more than five years. These websites sold substances that led to the fatal overdose of at least two individuals and fourteen non-fatal overdoses. These dealers now enter into the privacy of our own homes to entice and sell destruction to our children veiled under the illusion of being safe and legal. The formulation of analogues is like a drug dealer's magic trick meant to fool law enforcement. They didn't fool us and we must educate our children so they are not fooled either. Today's action will help prevent future deaths and overdoses, and will serve as notice for those dealing in designer drugs and the illegal use of the Internet.[13]

Tandy didn't give fuller details of the American Navy raves she briefly mentioned, where randy sailors were caught distributing and consuming a drug bought at one of the busted sites. The drug was 5-MeO-DIPT – a potent aphrodisiac. In common with many research chemicals, there has been little to no academic research carried out into the drugs' effects and it is impossible to say how this famously erotic compound works. But reams of online reports attest to its potency in the bedroom. Shulgin reported in *TIHKAL*: '(with 7 mg, orally) In one hour I was in a marvelous, sexy place. Everything was shaded with eroticism. Sex was explosive.'[14]

Operation Web Tryp was followed by other arrests in the US. In 2005, the owner of Pondman.nu, fifty-two-year-old David Linder of Bullhead City, Arizona, was jailed for 410 years, because a court found him responsible for the death of an eighteen-year-old man, Phillip Conklin in Hancock, New York, who took an overdose of another drug, AMT. Linder, known online as Dr Benway, was also much criticized by many of his customers for sloppy packaging and dubious business practices.

Operation Web Tryp was the first time the US had targeted drug dealing on the internet, and the first time the research chemical scene got coverage in serious newspapers and magazines. But despite the seriousness with which it regarded the crimes, the US did not judge the threat level correctly and therefore did not change its drug laws. It left the Analog Act in place and did not add many new compounds to the schedules. The fuzzy edges of that law meant that hundreds of drugs existed in a legal limbo and the customers were not an easy target for prosecution.

Only in the UK, where these drugs were completely and specifically illegal, was the decision made to target users. A few months after Operation Web Tryp, on 7 December 2004 officers involved in its British counterpart, Operation Ismene, arrested twenty-two people for purchasing the hallucinogenic drug 2C-I, again a Shulgin compound, from the companies busted in the US. Dozens of police officers from fourteen counties stormed people's homes for the crime of importing small personal doses of the ultra-rare psychedelics which, thanks to the 1999 law change, were illegal. One of those arrested in the UK action recalls 'the day of infamy': 'December 7th – it was like bloody Pearl Harbor! It was early one morning when they came for me. I was sleeping in and I answered the door bollock naked – I'd got stoned the night before. Next thing I know I've got a load of police in my flat, telling me to get dressed because they had a woman police constable there with them. A few months previously, I'd bought some 2C-I online, a tiny bit, from an American research chemicals vendor. But I thought the cops were looking for my hash, so I just co-operated, and handed them my stash tin. I had about 80 mg of 2C-I left. It seemed best to just be polite to them, I handed it over with my hash. They arrested me, but no charges were brought. I think the UK operation was carried out in order to appease the American Drug Enforcement Agency,' John told me.*

The boss of Britain's now-defunct National Crime Squad, Jim Gamble, who handled the raids, announced the events to the BBC. 'The internet has become the street corner for many drug users. By working in partnership with the DEA, the Crime Squad and police forces throughout the country have been able to arrest people suspected of purchasing drugs online. A drug supply route between the USA and the UK has been dismantled,'[15] he thundered.

The 'drug supply route' he claimed to have destroyed was actually a loose, unconnected group of twenty-two people making private purchases of extremely rare phenethylamines mainly for their own use. No one was jailed, and most were released without charge. It was hardly the crime of the century, and nor was it the investigation of the century – the customers of the websites had all paid for the chemicals using their credit cards. The DEA had emailed their names and addresses to the British police, who simply knocked on their doors or smashed them open with battering rams.

The British police did not monitor the online designer drugs market after that raid, and trained no officers in web surveillance techniques specific to drugs. Meantime, chemists and users' knowledge of the law, and of the new drugs available, were growing faster than the police could keep up with.

'The research chemical scene was an underground thing at first, there were only twenty-two of us in the whole UK who

---

* Not his real name.

got nicked,' says John. 'But it really started to take off after that. I think it was Operation Ismene that brought it to the attention of many more people. It was all over the papers. People who'd never thought of buying drugs online thought: "Oh, great, look – you can buy drugs online!"'

Over the next few years the unintended consequences of those raids were to crystallize. The police had wanted to clamp down on the trade, but had only encouraged its proliferation. And with the number of people using the web growing and connection speeds getting faster every year, they would soon be running a slow second place to the frontrunners in the field.

# 5

## *The Calm Before the Storm, and a Curious Drought*

After the Web Tryp and Ismene busts in the UK and the US, the global research chemical scene hunkered down and went even deeper underground. Many users commented on forums that the people involved in this drug culture were responsible adults who were being made to pay the price for the greed and indiscretion of the suppliers. Liberty was now clearly at stake for some people, and the net drug culture was on the radar of the authorities on both sides of the Atlantic. There was less online chatter about the drugs and their effects, as an informal code of silence settled like the dust after a stampede, but the scene continued happily enough.

Users like Benny, a pseudonymous 'thirty-something office worker' were by late 2004 discovering the joys of *PIHKAL* for themselves. Benny, who prefers not to reveal his nationality, was finishing university and was winding down his commitments at NGOs, where he volunteered and worked as a political activist, and now had time on his hands. He had always been fascinated by psychoactive drugs, but he was living in a backwater where there were no good drugs available, so he went online and ordered a mushroom growkit from a Dutch smartshop. He successfully grew mushrooms, was surprised by how easy it was, and said his first trip, in retrospect, was almost an overdose. 'It changed me. Irrevocably. From then on, psychedelics became part of my life,' he told me.

He, like many other research chemical users, wanted to use these drugs as a means to self-exploration and self analysis: 'Psychedelics disassembled reality and showed me a new way of looking at basic concepts like time, space, the concept of self and the distance we have from other people; or the construct of one's own personality, and the way we manipulate and write and rewrite the narrative that we think is our life. This was fascinating – to take a step back and rethink those concepts that I'd taken for granted all my life. After coming back from the first trip, a few things were clear: first, psychedelics would gradually change me. And I was keenly aware that the changes would be beneficial, that they were benign in nature; I was sure of it so I pursued it. In life, you're battling through the undergrowth and every so often it's good to climb a tall tree to get your bearings – this is what psychedelics do for me.'

Benny started reading obsessively about the psychedelic experience, and soon came upon Shulgin's work, which was by then available free online to anyone with a cheap home computer and an equally inexpensive modem.

'With friends I had made on Usenet, I found that many of those substances were apparently readily available from fine chemical companies,' he said. 'I then wrote a lengthy and completely fake research proposal to a venerable chemical company, a reputable and viable operation which caters to universities. They were happy to open an account for me and I ordered some 2C-I.' This is the drug many UK residents were arrested for in the Operation Ismene busts. 2C-I is a ring-substituted phenethylamine, a classically psychedelic drug that enhances colours and gives users a sharp energy jolt. It is active at around 10 mg, and one of Shulgin's group of human experimenters recorded a positive experience of the drug in *PIHKAL* entry #33:

(with 16 mg) The 16 was a bit much, I realized, because my body was not sure of what to do with all the energy. Next time I'll try 14 or 15. However, my conversations were extremely clear and insightful. The degree of honesty was incredible. I was not afraid to say anything to anyone. Felt really good about myself. Very centered, in fact. A bit tired at day's end. Early bedtime.[1]

Benny found a firm in Switzerland that supplied the compound. 'I don't even know why this firm would carry this chemical,' he said. 'They usually carried pharmaceutical intermediates, but this one stood out and I ordered it . . . I found it so ridiculous when I received it . . . How could someone allow me to order powerful drugs from the internet and receive it via regular mail? It was inconceivable – and ridiculously easy. This was long before [today's sites] came along and made it as easy as ordering books from Amazon.'

In common with many other psychonauts, Benny soon perfected his skills at unearthing sources for the drugs found in *PIHKAL* and *TIHKAL*, using Google to search the new large chemical directories that came online at this point. If you simply type in a number identifying the chemical, you receive a list of companies that can provide it, with price quotes. 'I'd compile huge lists of them and go through every one of them and try to see if they had anything interesting and to see if they would be prepared to send it to me,' said Benny, who became something of a psychedelic completist. Pioneers like this were at this time acting in relative isolation; newsgroups and bulletin boards were the preferred hangouts, but within a few years as the technology became more transparent and web use became ever greater, such actions would become much more common. The entire process of researching and buying new designer drugs was becoming even more commercialized and commodified as internet use proliferated. For some people in the twenty-first century, taking research chemicals became as normalized as taking Ecstasy had been for their counterparts just a few years previously.

In around 2004, net speeds in Europe changed as broadband connections became widely available, and the continent started to play catch-up with the US, where faster connections had been commonplace for a few years. Telephone firms' monopolies were being dismantled, their services packaged out, and bidding invited from the private sector. Customers started to pay flat subscription fees for constant access to the internet, rather than paying per minute to dial in, and the net became firmly embedded into everyone's daily lives in a way that had never been seen before. Previously, modem speeds had been limited to 56K; broadband connections increased the speed by a factor of ten overnight, to 0.5MB. That doubled within a year.

Faster speeds meant net users spent more and more time online as the experience grew less frustrating, and network effects – whereby the service became more useful as it became more populated – started to be felt. There was a corresponding increase in the use of the web for people with niche interests, who could now coalesce and communicate in ways that previously were available only to those with the skill to configure Usenet servers and subscriptions. Thousands of online communities were formed by people using the newly popular and much simpler bulletin board systems, such as those released by vBulletin. Some were open source, meaning that owners could tweak the boards' look and feel and functionality easily, and that they were easier to technically configure than any previous web community software.

Some of these forums were dedicated to the discussion of psychoactive drugs, indeed all drugs, from opiates and psychedelics to MDMA and research chemicals. There were sites discussing the most specific and niche of psychoactive interests, from DMT extraction to magic mushroom cultivation, from ways to potentiate a heroin hit to the most efficient ways to grow marijuana or make crack cocaine out of powder cocaine. There was a similar atmosphere to the early alt.newsgroup days, but the scene was much more open and accessible to non-expert users, and much more heavily populated. Whereas Usenet groups might involve a few score emails a day, now thousands of conversations were carried out in virtual real time across hundreds of topics. The online drug culture was becoming more established and more mainstream; this was a cultural as well as a technical shift.

The Bluelight bulletin board launched in its earliest iteration in 1997 as a small community for American MDMA users. It grew in size and influence following Operation Web Tryp, as more people became aware of the possibility of buying drugs online and net use became more prevalent, and is today one of the net's busiest communities of drug-users, with over four million posts on 200,000 threads. There are 195,000 registered users, and the site gets over one million visitors each month and serves over three and a half million pages in the same period. In the last year, it has had 10,967,685 visitors and served almost fifty-five million pages, demonstrating that the site is read by many who are casual visitors rather than registered members. It is a one-stop shop for anyone who is researching drugs and who demands detail and accuracy – something sorely lacking in most media reports and official discourse. The popularity of the site is one of the key measures of the popularity of the online drug scene. It now rivals Erowid as a place to learn about research chemicals and other drugs, though it is far less formal and respectable.

Although Bluelight is a freewheeling and anarchic place, it has an avowed goal of harm reduction, a drug strategy that is now commonplace in countries such as the UK. Harm reduction advocates would argue that as long as people are going to take drugs, and they remain available, and jail capacity and public funds are not infinite, the most rational response is to reduce the amount of harm drugs cause users. This, officially, might mean offering advice on how to inject heroin hygienically, or offering needle exchange programmes. On the web, it could include information on the steps you might take in order to diminish an MDMA comedown, or advice on making and smoking crack cocaine less harmfully. The concept, naturally, is rejected by the less liberal, and by those ignorant of the realities of drug use in the twenty-first century.

Bluelight's most rigorously enforced rule is: no sourcing. Ask for a source for drugs or research chemicals and your threads will be deleted. It's an understandable position to take when the substances under discussion are either illegal or extremely potent and can harm inexperienced or ignorant users.

The site is well moderated by those expert in each field of the subforums, which cover hundreds of drugs and other areas of interest. The users there range from the supremely expert to the inarticulate to the terrified or addicted. But generally Bluelighters are at least a couple of years ahead of the curve when it comes to new drug trends, and some of them far more informed than many doctors, police, or policymakers, and the site is often approached by radical academics looking for assistance with cutting-edge research into users' changing habits. It is both the frontline of an avant-garde web drug culture, and the first place any informed drug user or academic looks when new chemicals appear on the international markets, especially the online market.

Many posters share their encyclopedic, often professional or academic knowledge freely, collaborating to advance knowledge of harmful combinations or to explain dangerous metabolic processes that the body exerts on drugs. For example, when cocaine is taken with alcohol, the liver bonds the two molecules – cocaine and ethanol – to produce cocaethylene, a more toxic and longer-lasting poison than either alone. Advice such as this from expert users with knowledge of biochemistry sits happily alongside the confused ramblings of newcomers. Experts also explain which combinations of drugs lead to better or more intense experiences, and PhD-level knowledge of bio-pharmacology and organic chemistry is sometimes required to decipher some of the information. Some vendors worldwide have extracted information on new drugs from posts written by the site's expert users, then synthesized the compounds and sold them on the web.

Some users on the site are themselves chemists, or know chemists who make new drugs in home laboratories, much like latter-day Shulgins. Synthesis discussion is banned on Blue-light, but naming novel compounds and reporting on your experiences with them is not, so vendors need only cut and paste the chemical names from these trip reports, email them to labs in China or Eastern Europe, commission a custom synthesis, and then import and sell them.

Again, Ecstasy had a starring role in the genesis and development of this online hangout. 'Bluelight began . . . as a small message board hosted on Bluelight.net called the MDMA Clearinghouse,' said Alasdair Manson, a forum moderator. There, people gathered to accumulate every scrap of information on how to take Ecstasy safely and enjoyably, the web and users filling the void of official inaction. 'It had a small number of regular participants and was a close community. It was suggested that a more permanent home be created so the board could be stable and the content could be better managed and stored more permanently,' Manson told me.

The site today is well organized into subfora, with support groups for people addicted or suffering from problematic use. Each forum is managed fairly by voluntary moderators and discussions are, in the main, polite affairs. I asked Manson how he felt about the fact that people might read Bluelight and harm themselves as a result of information they found there. Any journalist with malicious intentions could scoop dozens of stories and quotes from the site to use as evidence that it encourages damaging behaviour. He told me, 'It's inevitable that, given the very nature of the subject matter, people will learn about things they didn't know if they had never visited Bluelight. Could one person learn something here which might ultimately harm them? Sure, but users also have to take responsibility and that's a common theme in discussion at Bluelight. I do not believe we're enablers in the broader picture, and perhaps the single most obvious example of that in practice is our rule about "no sources". I'm proud of the fact that, ten years after I joined, Bluelight is still going strong and doing its job. If anything, I'm most proud of the fact that Bluelight is very much a product of its participants. A core of dedicated individuals (staff and participants) has donated time and money to keeping it running. People use Bluelight because they can't get the information it provides anywhere else. The world is a free market of ideas now, and the internet is an enabling tool the scale and shape of which we're only just beginning to understand.'

The site is far more trusted by drug users than official services such as the UK's much-derided Talk to Frank drugs helpline. It's a difficult truth for the government to accept, but people are taking more drugs than the authorities even know exist, in ways they would not believe, and it is sites such as Bluelight that keep them safe. Another Bluelight moderator is clear on why official services can only ever fail to achieve their goals. 'People don't take government services seriously; it's the official tone that's the problem,' he told me.

That official tone can be more than high-handed – it can be dangerous. 'Someone I'm very close to rang the government's Talk to Frank [helpline] when they were having a panic attack after overdoing a few substances and booze,' said one poster on the drugs subforum of Urban75, another busy bulletin board based in Brixton, south London. 'Frank told them they were very stupid and that mixing drugs and alcohol can lead to brain damage and death. Really cool thing to say to someone who's having a panic attack.' Mike Slocombe, Urban75's owner, told me, 'People are going to take drugs no matter what the government think. On Urban75, we give honest advice, like an older brother who's been there, done that.'

From 2004 onwards, the web was changing the way people swapped information and bought and sold goods. Web use started to shift inexorably away from a passive model of consumption into a more active model of participation. This was the beginning of the Web 2.0 era, though it wouldn't be called that for a few years. It became possible for complete technical novices to self-publish web content using simple blogging software, doing away with the need for any skills in HTML or other coding languages. Free webhosts launched, supported by advertisements, giving users the ability to share and disseminate their ideas and thoughts at no cost.

Wikipedia, launched in January 2001, was arguably the first widespread Web 2.0 technology, for which users supplied content using their own expertise and drawing on their own interests and obsessions. At first the content was as flaky and patchwork as an online encyclopedia created by untrained volunteers might tend to be. Today, though, the site has twenty-three million articles, 4,079,607 of them in English, and grows by the day, as volunteers, or Wikipedians, mass-collaborate to update, create and revise the pages almost instantly. (By comparison, the *Encyclopedia Britannica*, which halted printing in 2012, had just 85,000 articles.) Wikipedia soon became an essential resource for those looking for information about new drugs, and the site started publishing entries on the drugs along with their Chemical Abstracts Service (CAS) numbers – the unique identifying code that serves as a chemical Dewey Decimal system. With this information, would-be vendors or users could easily find chemical firms that carried the compounds they wanted.

In 2003 MySpace opened the web to a whole generation of teenagers, to whom the concepts of packet-switching were as alien as the concept of not using the net as their first port of call for entertainment or communication. The site, with its super-vernacular design, clumsy layouts and clashing colours, was as riotous and impenetrable as any poster-adorned bedroom wall of previous eras.

It wasn't until late 2004 that the phrase Web 2.0 was officially coined by technologist Tim O'Reilly, who correctly identified the future of the web – it would become a model driven by user-generated content, mass collaboration, global sharing and cross-border participation.[2] That year, The Facebook, a web version of the college yearbook popular at American universities, launched. By 2008, it had lost its definite article and gained 100 million users; just four years later, in 2012, a figure comparable to the entire population of India – 955 million people – had an account and the site became many people's primary locus of online activity. Every redesign of Facebook's software was conceived and executed to encourage more and deeper sharing of content, to open and ease the flow of information from one person to another. Users have happily, if unknowingly, delivered millions of gigabytes of valuable data to firms whose only specialism is to analyse likely ways to use that data to sell users things. Within a few years, consumer goods would have their own pages, and people would befriend these brands.

In 2005, YouTube came online and video sharing soon became not just a possibility, but a cultural norm. Today forty-eight hours of video are uploaded every minute, as handheld cameras have given way to smartphones and net users can add content on the move. Today, almost eight years of content are uploaded every day. The following year Twitter launched and was initially derided as a passing fad, a super-condensed, super-trivial microblogging platform for narcissists. Now, it is an essential tool for its 140 million members, who include politicians, brand specialists, marketeers and journalists.

The launch of the iPhone in 2007, like the Macintosh in 1984 and the iMac in 1998, changed the way people used and made the net. Now, the web was mobile. Today, Apple alone has sold 315 million mobile internet devices. Hundreds more types of smartphones were launched in the following year. The western world wasn't just online, it was always online, and producing and sharing content and information whose nature is human, but whose scale is beyond all human understanding. In 2000, there were 361 million people online. By the end of 2011, that figure had increased to 2.27 billion, with the majority of them, over one billion, in Asia. The number of web users in Europe has quintupled in the last decade in Europe, and doubled in North America. Overall, internet connections worldwide increased by 528 per cent between 2000 and 2011. Nearly everybody in all countries that had electricity in the northern hemisphere, in the rich countries, in the countries where boredom could be measured and avoided in the flickering of a synapse – in the drug-consuming countries – had net connections. The cultural shift away from suspicion and confusion over the net had been replaced by one of complete human-network symbiosis.

By 2004 most net users were comfortable making purchases online. Amazon had gathered millions of credit card details and delivery addresses for its customers, and had started to increase its offering beyond books. It had also opened its marketplace to hundreds of thousands of other vendors, and allowed them an unbranded version of its highly complex ecommerce system. Suddenly every retailer in the UK and US had the option of selling their goods online easily. Meantime, Apple's music service iTunes had harvested the credit card details of all its users and made the experience of buying music as simple as point and click. Online payment processors such as PayPal were now trusted and used by millions of people. Competitors and alternatives started gaining traction; among them Canadian firm AlertPay.

That growth soon became deep-rooted: the Interactive Media in Retail Group said in a May 2011 report that business-to-consumer global ecommerce sales in 2010 amounted to 591 billion euros – a twenty-five per cent year-on-year increase, with the trillion euro mark set to be broken in 2013.[3]

One of the first moments of mass online drug purchasing happened in the UK in 2004. Information started to come online at about this time that clarified a long-confused area of British drug law. Britain's indigenous strain of hallucinogenic magic mushrooms, *Psilocybe semilanceata*, or liberty caps, with their distinctive nipple-shaped domes, are powerful, natural psychedelics, and grow only in the early autumn before the winter frosts set in (liberty caps are notoriously fragile and cannot easily be raised artificially or indoors). All mushrooms are the fruiting crop of a much larger organism, the mycelium, which lives underground and survives many years. This means that fields where the fungi appear annually will deliver the goods year after year, but their whereabouts is often not shared. These mushrooms grow in huge abundance in many areas of the UK, and since the 1970s free festival scene and the post-LSD growth in interest in psychedelics they have been a regular feature on the psychedelic cognoscenti's calendar.

Until 2003, most people in the UK who used them knew the law: mushrooms were legal in their natural state, and you could only be prosecuted for having them if they were dried, frozen or cooked – that is, if they had been prepared for consumption. Debate among the sarcastic and the pedantic was split as to whether human beings lying on the ground and grazing on them, as other animal species might naturally do, were actually committing a crime.

But it was well known that the Dutch allowed their citizens to buy artificially cultivated psychedelic mushrooms. These were sold fresh, year round in small punnets in shops that catered to the local market as well as curious visitors to Amsterdam. Might the web and the new technologies of ecommerce be able to bridge the gap? The answer came in 2003, when Home Office official Ian Breadmore sent an extraordinarily, if inadvertently, helpful letter to a headshop owner who had inquired as to the legality of fresh magic mushrooms in the UK. Breadmore told the inquirer with brilliantly British bureaucratic efficiency that it was not illegal to grow and pick psilocybin mushrooms and eat them fresh; nor was it illegal to sell or give away a growing kit containing mycelium, and nor was it illegal to sell or give away a freshly picked mushroom that had not been prepared in any way.[4]

An online market began and grew rapidly from 2004 onwards, and there were dozens of sites selling the mushrooms fresh in polystyrene punnets via mail order all over the UK. These websites also sold simple growkits that would yield hundreds of grams of fresh mushrooms – enough for several people to hallucinate a weekend away – costing only ten to twenty pounds. It was inevitable that offline retailers would soon want a piece of the action, and soon every high street bong vendor and online headshop in the land got in on the act and started selling the fresh mushrooms and the growkits.

These new mushrooms looked nothing like the spindly liberty cap: they were Mexican breeds, huge and chunky specimens of *Psilocybe cubensis*, descendants of samples harvested from fields of cow dung in Mexico and Cambodia, initially grown in Holland and sent by chilled express freight into Heathrow daily. The net had facilitated the exchange of expertise and information, and brought together consumers and producers across the seas. Later the UK market supplied itself. You could buy magic mushrooms legally in dozens of British cities. On Oxford Street in central London, shops sold them blatantly, advertising their wares with psychedelic hoardings in the street. By summer 2005 they were on sale in chic market stalls in wealthy London suburbs. Festival stands sold out within hours of opening, while eBay vendors did a roaring trade. At free parties the fungi were passed around daintily by dreadlocked ravers like bizarre hors d'oeuvres as the sun rose. Mushrooms were being eaten openly at gigs and parties and barbecues and raves as for a few brief months the government dallied and prevaricated.

Just before Parliament went into recess in August, it closed the legal loophole that had been exploited by vendors in a 'wash-up' session, where outstanding legislation is quickly written into law, and mushroom vending became illegal.

The mushroom craze demonstrated that as soon as new drugs became available, especially if they were legal, British people would take them. What's more, they were happy to buy their drugs online in the same way they now bought books from Amazon.

Fresh magic mushrooms were banned in Holland in December 2008 too, after a seventeen-year-old French girl, Gaelle Caroff, jumped to her death from a bridge whilst visiting the city on a school trip. She had a history of psychiatric problems, the Associated Press reported. Paul Van den Berg, who worked at a shop selling the mushrooms, told the *Daily Telegraph*, 'It's all the fault of tourists, especially the Brits. They misuse alcohol at home and come over here to do the same with hash and the so-called "magic mushrooms".'[5]

Dutch ingenuity quickly supplied an alternative: psychoactive sclerotia, or truffles, were soon bred for mass-market sale (they had been available previously, but the ban sparked a boom in their cultivation). These truffles offered exactly the same effects as the banned product, and remain completely legal at the time of writing in Holland. They are illegal in the UK, but for a few hundred pounds many vendors in Amsterdam will happily and illegally send a kilo or less by Fedex or UPS or DHL to most countries in the world.

It is an indisputable fact that millions of British people spend a good deal of their disposable income – and free time – using illegal drugs. Pick any year and you'll find that the UK is excelling at some form of drug use in one way or another, and the nation also has a parallel and serious problem with binge-drinking. Working out why the British use intoxicating substances to such excess keeps hundreds of civil servants, politicians and academics busy (and funded), exercises our police force and helps them justify calls for ever-increasing budgets to tackle the issue.

In around 2008, it became clear both from police seizures and from anecdotal evidence among users both online and in real life, that even as use and consumption was rising, the quality of most drugs sold in the UK was declining, especially that of cocaine and Ecstasy. Cocaine quality had been steadily falling for the seven years before the 2011 visit I made to the Forensic Science Service (FSS) labs in Lambeth, south London. As you enter the drug-testing room at FSS, the smells slap you in the face. There's a jungle reek of high-grade marijuana; a saccharine-sweet solvent odour wafting from sacks of yellow amphetamine; a kerosene tang from large bags of cocaine strewn casually on steel-topped tables. The white-coated scientists who move among the aromas appear oblivious, however and quietly get on with methodically weighing, measuring and documenting their findings. A huge clear bin liner a few benches away is filled with twisted, knotted lumps of base amphetamine, gnarled like melted cinder toffee. A sack of spongy-fresh skunk cannabis overflows, the pounds of tightly trimmed top flowers adding an earthy and pungent fragrance to the olfactory overload.

Many police forces across the UK until recently used the lab to measure the purity – or impurity – of drugs samples seized in raids and arrests. This nondescript room at one end of a bland, municipal 1970s office corridor can tell us much about social, criminal and chemical trends in the UK. 'We have a lot of work; there's never any shortage of cocaine. But quality has slipped quite markedly in recent years,' said Dean Ames, the FSS's drugs intelligence advisor.

Through the 1970s and 1980s cocaine was synonymous with money and status. Now, though, the drug is to be found in even the smallest towns and villages; cocaine, police say, has penetrated every corner of Britain. As use has increased, so the price and purity of the drug have fallen dramatically, a situation few could have predicted even ten years ago. 'A typical importation quality at the start of the decade would be around seventy per cent pure, and a police seizure, at distribution or street level, would be around forty to fifty per cent,' says Ames. 'We've seen a reduction in the quality of import cocaine, which is now at about sixty-five per cent, and the quality of the drug on the street is, at best, down to twenty or thirty per cent. It can be as low as ten or even five per cent. In some cities, such as Liverpool, there is sometimes no cocaine found in samples at all.' Instead, he said, they will contain a mix of caffeine, benzocaine and paracetamol.

Kelly Burt, a twenty-three-year-old assistant forensic scientist, focuses intently as she weighs out a tiny quantity of what is presumed to be cocaine using a scale that measures to one ten-thousandth of a gram. She has worked here since graduating in forensic science and investigative analysis from Kingston University, and is now a drug analyst and heroin profiler. Can she judge cocaine purity on sight? 'No, I'm always surprised. I can tell cocaine by the smell of it now; it's very distinctive,' she says. 'But as for purity, no. Sometimes I think it's really pure and it turns out to be just one per cent. There's no way of knowing without using the procedures we do here.' Most cocaine she has seen in the last five years was of very low quality, she said. 'It's generally below ten per cent, the street deals. Nine times out of ten. Most supposed grams weigh very little – as low as 200 mg [one-fifth of a gram]. Only rarely, when we get a larger seizure, do we see any samples with higher purity than that.'

This decline in quality in the UK was caused by a number of factors. An interesting one is the pound's collapse against the dollar and the euro – the cocaine currencies. In May 2007, the dollar was trading at around fifty pence. Today it's worth about sixty-two pence – meaning it costs around twenty-five per cent more in sterling to buy the same amount of product.

Matthew Atha, director of the Independent Drug Monitoring Unit, which acts as an expert witness in British drug court cases, told me in 2011 that most cocaine in the UK was, by then, ineffective. 'It's pretty pointless trying to use street cocaine as a stimulant. The amount you need to take is so large, people would be better off with a cappuccino these days, to be honest. It's much better value for money.'

The import and sale of benzocaine, a numbing agent used legitimately as a topical anesthetic, also contributed to the decline in quality. It is used covertly to cut down the powder and bulk it out, since users think numb gums mean a high-quality product. Snorting it will make your nostrils and then your teeth numb, but the drug does not trigger the same flood of dopamine that true cocaine does. Likewise, it prompts no alertness, so often dealers simply sell benzocaine mixed with caffeine.

David John Wain, from Hayes in London, imported more than seventeen tonnes of cutting agents, including 7,000 kg of benzocaine, from China. This was almost as much benzocaine as the whole legitimate UK market used in a year. He was jailed in 2010 for twelve years, as part of a new Serious Organised Crime Agency (SOCA) strategy called Operation Kitley, which targets those supplying chemicals to the cocaine trade. SOCA senior investigator Trevor Symes told the BBC: 'Even after we warned him, he [brought] in more tonnes of chemicals. He felt the law did not apply to him. He knew, and he knew full well, that he was importing chemicals for drugs on a massive scale.'[6] Wain was prosecuted under the Serious Crime Act 2007, which targets those who help gangs but may not be involved in the end offence.

The British Crime Survey in 2008 put the number of cocaine users in the UK at 974,000[7] – more than any other country in Europe. By 2011, Britain had been Europe's top cocaine-consuming nation for the previous six years. So there was, in 2008 in the UK, a huge and generally unsatisfied market for an affordable and effective stimulant.

A similar thing happened to the MDMA markets in 2007–8. After years of high-quality MDMA, the British and European market suddenly dried up. Britain's supply was among the world's worst, thanks to its geographic isolation as an island. But Europe, too, was suffering a mysterious MDMA drought.

In the 1990s, when Ecstasy quality first dropped briefly, news spread slowly. In 2007–8, however, Web 2.0 technologies such as forums, together with the online sale of Ecstasy pill-testing kits, meant that people were forewarned and able to communicate the dangers publicly, and instantly.

Pillreports.com is an unofficial barometer of the Ecstasy trade, especially in Europe. In common with many other online drugs communities, the site hosts user reports and images of pills bought as Ecstasy, with dates and test centre reports, if the buyer is lucky enough to live in a country such as the Netherlands with free testing of pills. Otherwise users will test the pills themselves, using pill-testing kits containing a chemical known as the Marquis reagent. To use them you simply crumble a chip of a pill onto a plate and add a drop of the liquid; if it goes black the pill contains a primary amine, likely an MDMA-like substance. A colour-coded chart offers users some way of telling if the pills are poisonous or safe.

The site, which was set up in 2000, offers a fascinating window into a hidden culture, and plots geographically and publicly the quality of drugs across Europe in real time, while most official reports lag by at least a year. The site's servers are located in the Netherlands, thanks to the country's lenient and liberal drug laws. It has 29,345 user-generated reports on pills sold as Ecstasy, and gets around 15,000 unique visitors each day. It is a self-managing community, with moderators keeping an eye on threads to ensure that dealers do not use the service to advertise their products. 'Like any site that crowdsources user reviews there is always the danger of people gaming the system, but we find that good information always drives out bad,' the site's administrator told me by email. To those who argue that the site encourages drug use, he offers the baldly factual response: 'Studies on the influence of drug information, and particularly pill testing, have shown that the more information people have about what is in pills the less likely they are to take them.'

Pillreports proved its worth most significantly in late 2008 and throughout 2009, when its pages were covered in pink reports – highlighted in that colour to show that the drugs bought contained no MDMA. There was evidently a worldwide drought of the substance; its crystal form had not been seen in months, and considering the size of the UK market, which consumed millions of pills per year, something was clearly amiss. Nobody could understand why this was happening. All over the net people were reporting that they were sick and tired of buying tablets that made them ill, and in clubs across the UK users were reporting headaches, fever-like symptoms, unpleasant comedowns and bizarre hallucinations.

Health professionals and toxicologists worked out that users were actually eating a class of drug known as piperazines. These were legal compounds that had never been abused recreationally, which had seldom been seized at borders or on the streets, and which had never been considered by the ACMD as being dangerous. To the uninitiated, piperazines might have passed for a dose of MDMA, but most users rejected them, even though they were being sold in some cases for just a couple of pounds each, because they offered a brief, mind-mangling high.

Some piperazines are used as anti-worming agents in dogs, and some in medical research centres in tests for the migraine headache industry, to induce migraines in test patients. A side-effect of these piperazines was nausea, and analysis of pills from 2008 showed that some criminal chemists had even included a small dose of domperidone, an anti-emetic drug used to manage vomiting in pregnant women suffering morning sickness, to mask it. Why were the crime syndicates using these foul concoctions? Surely no dealer would want to kill or sicken his customers, or worse, deter them from spending more money? The Ecstasy market started to collapse.

Unbeknown to European users, in Cambodia between 2007 and 2008, a series of events had occurred that link together every strand of this story. This was the fulcrum on which the events of the next few years would tip.

# 6

## *Mephedrone Madness: the Underground Hits the High Street*

In the pristine and remote Cardamom mountains in south-western Cambodia in 2008, local forest rangers and conservationists from Flora and Fauna International, a British NGO, came upon a smoking, abandoned campsite. The scene was a bizarre mix of charnel house and war movie, but instead of piled corpses and bones, there lay scattered the husks of a tree beside a vast cauldron over the embers of a large fire. Clothes, books, food and mobile phones were strewn around and a roasting animal still cooking on the spit of a camp fire spoke of recent occupation, and flight. An AK-47 bullet casing lay on the ground near screw-cap plastic gallon containers of a sweet-smelling oil.

The oil was safrole, distilled from recently felled trees that had stood for several hundred years before the arrival of the poachers who had set up the camp. They had cut a savage path into the heart of this, the longest contiguous tract of virgin rainforest in South-east Asia, in search of the *mreah prew phnom* tree, whose bark and roots contain unusually high concentrations of the oil.

In this and other ramshackle safrole factories, the essential oil is distilled from the tree's bark and roots by first shredding them with mechanical strimmers and grinders, and then suspending the stripped material over heated water in a vast iron cauldron, which has an outlet pipe in its top that directs the now safrole-rich steam into a condensing chamber. In this form – in which it has no psychoactive effects – it has long been used in traditional Khmer medicines. But the oil, if it had been taken to the capital, could have been sold for fifty dollars a litre. One litre of this oil can make anything up to 1,000 Ecstasy pills.

In late 2008, in Pursat, 170 km west of the Cambodian capital Phnom Penh, UN anti-drug officers destroyed 33 tonnes of safrole oil,which had been earmarked for use by drug gangs in the Netherlands to make Ecstasy following Shulgin's recipe.The oil, confiscated over the preceding weeks and months, had been produced in illegal safrole labs like the one in the Cardamom Mountains, that made an average of sixty litres a day. If the thirty-three tonnes of oil had got to Holland, it would have translated into as much Ecstasy as all the users in Britain alone combined would have normally taken in five years. Instead, it was burnt.[1]

There are many ways to make MDMA, but safrole is the simplest synthetic route; the chemistry of the two molecules is not very different, as can been seen in these diagrams:

*Safrole*

*MDMA*

More importantly, the steps from safrole to MDMA are fewer than by most other methods and the yield is often higher. The impact of the events in Cambodia was therefore profound and lasting – and it inadvertently caused the appearance of the drug that came to be known as mephedrone.

An analysis of the data published by the United Nations Office on Drugs and Crime (UNODC) in the World Drug Report 2011 stated that in 2004, global seizures of what the UN calls 'Ecstasy group' substances were 12.7 tonnes.[2] That figure dropped to 9.7 tonnes in 2005, stayed steady for 2006, and then in 2007, almost doubled, to 16.5 tonnes. But seizures then dropped vertiginously to 6 tonnes in 2008 and 5.4 tonnes in 2009. These were the two years of the global MDMA drought and include the year of the 33-tonne Pursat oil burn that took out the precursor required to synthesize 260 million pills. If we calculate 245 million pills at a rather generous 125 mg each tablet, that would add up to around 29 tonnes of MDMA – enough to starve the world market for all of 2007, 2008 and part of 2009.

Not only were the precursors getting harder to find, but in June 2007 the world's biggest Ecstasy bust had taken place in Australia. The global MDMA market is mainly controlled by Russian and Israeli organized crime groups, Dutch-born chemists and Israeli and Italian smugglers. An alliance of Italian crime bosses and motorcycle gangs in Australia arranged the import of fifteen million pills in tomato cans from Naples to Melbourne in 2008. They were caught after 100,000 telephone intercepts and thousands of hours of surveillance in an investigation that involved 800 police. In May 2012, Pasquale Barbaro, a fifty-year-old farmer from New South Wales, and a member of the powerful Calabrian 'Ndrangheta crime syndicate, was jailed for life for the offence. His accomplice, fifty-five-year-old Saverio Zirilli, was jailed for twenty-six years.

In late 2008 and early 2009, then, users were crying out for either MDMA or some new replacement. There was a huge untapped and unsatisfied market, filled only with piperazines. A person with close links to those involved in the top-levels of global MDMA manufacture revealed to me what prompted the emergence of piperazines and other adulterants as a response to the increase in prices for safrole and other precursors between 2006–2009: 'Precursor prices [for compounds such as safrole] had gone up considerably from 2006 onwards. The steadily increasing security in European ports was making it harder to bring bulk quantities of precursors in, even when you could find them. This was sending the price of manufacture up, not by a huge amount yet but it was concerning manufacturers,' he explained. 'A number of international syndicates took a vote on whether or not to raise the wholesale prices on pills or look for cheaper substitutions. Apparently the vote came down on the side of substitutions and it was after this that we started seeing more and more pills with what we would consider adulterants, but which the industry was trying to promote as substitutions.'

But the strategy backfired, when users became wary to the point that many of them stopped purchasing the new pills that sickened them so badly.

Around five years earlier, an underground chemist known as Kinetic had posted the web's first synthesis of a new drug to the Hive. 'I've been bored over the last couple of days,' he wrote in 2003, 'and had a few fun reagents lying around, so I thought I'd try and make some 1-(4-methylphenyl)-2-methylaminopro-panone hydrochloride, or 4-methylmethcathinone as I suppose it would be commonly called.'[3] He went on to document how he had synthesized 4.8 g of the drug in forty-eight hours using toluene, a simple solvent, as the precursor. He then, like Shulgin before him, tried the drug he had made, and reported back to a fascinated audience of fellow renegade chemists. But this drug was nowhere to be found in *PIHKAL* or *TIHKAL* – it was almost entirely unheard of, and certainly never seen before on the mainstream drugs markets. Kinetic had used his knowledge as a chemist to devise mephedrone, a new analogue of methcathinone – a powerful and illegal stimulant drug related to methamphetamine, or crystal meth. These diagrams show the similarity between the three compounds:

*Methcathinone*

*Methamphetamine (crystal meth)*

*Mephedrone*

Over the next few hundred words Kinetic laid out, in a remarkably concise yet exhaustive guide, the method, tools, chemical reagents and potential pitfalls involved in the manufacture of this new drug. He did this not to earn money, or to get a job, but to show off his mastery of organic chemistry, to gain kudos, to share his knowledge freely. It was certainly not his intention to trigger a worldwide drug craze, but within a few years, that's exactly what happened. Kinetic told how he had spent the day synthesizing the material, and a few hours after scraping out the 4.8 g of crystals from his reaction cylinder, had snorted 400 mg. His central nervous system rode the waves of powerful, if short-lived euphoria. He wrote:

> 400 mg was quite a lot to take in one evening, but it wasn't too long-lasting so I kept 'topping up'. The rushes after each line were amazing, and I remember feeling very much like I do when coming up on Ecstasy, but four times! That beautiful weak feeling, when you just think 'Oh, fuck, I feel so fucking good . . .' It's more euphoric than I remember my first time on the butane methcathinone analogue to be. I felt very compelled to do things, but I was completely unable to keep my concentration on literature searches I was trying to do at the time. I had a very strong urge to socialise, and almost went clubbing, but thought better of it. I could feel the rushes of energy coming across me, and after that, a fantastic sense of well-being that I haven't got from any drug before except my beloved Ecstasy.

Kinetic's new drug, which later became known as mephedrone, was indeed similar to Ecstasy, in that it made you euphoric, excited and energetic, and it provoked empathy and openness. Not only that, but mephedrone was easy to make and it was legal in most countries in the world, since no country had ever seen a methcathinone analogue before, and in the UK, certainly, there had never been a need to pre-emptively ban modifications around the basic methcathinone molecule. The drug's main difference to Ecstasy was its short duration – about forty-five minutes compared to the four- to five-hour Ecstasy experience. This would make it addictive – as soon as users came down, they wanted more.

On 5 April 2003, at 21.33 exactly, Kinetic posted another message: 'Oh, and I really just have to say a big "Fuck You" to the UK government and their stupid drug laws, since I'm high as a kite and there's nothing they can do.'

Kinetic was right: he was defiantly exploring a class of drugs, cathinones, analogues of which had never been controlled in British law, unlike the phenethylamines and tryptamines. And as there was no analogue ruling in UK – unlike America – that blanket-banned products and left legal doubt hanging over users and vendors, he simply rerouted around the law, safe in the knowledge that by obeying it, he could not be punished. Nor could those who followed his instructions, four years later, to create the legal drug that would fill the MDMA gap.

The consequences of this underground bulletin board posting were as unintended as they were unanticipated. Perhaps just a few dozen highly dedicated, highly educated scientists (and various 'lurkers' who listen in, but do not contribute to the discussion) were privy to Kinetic's quietly seismic message. But that post fired the starting gun on the current chemical arms race between clandestine chemists and governments worldwide. 'The designers of mephedrone stumbled across something with the pharmacology somewhere between cocaine and Ecstasy that was cheap, legal and freely available. And this coincided with a drop in the quality and availability of other drugs,' says John Ramsey, chief toxicologist at St George's, University of London.

The global MDMA drought caused the mephedrone pheno- menon in a pitch-perfect piece of substance displacement – if one drug disappears, another will replace it. Substance displacement would soon cause the new research chemical scene, now rebranded and introduced to the public as the legal highs scene, to grow at a rapid rate. The internet and the use of social media and new ecommerce techniques guaranteed its influence spread even faster. The way many drugs are bought, sold and taken in the UK and across the world was to change for ever.

At first, mephedrone was hidden inside branded products rather than being sold as a chemical compound in its own right. In early 2007, an Australian subforum on the Bluelight bulletin board was buzzing over the release of branded legal highs named NeoDoves and SubCoca. They were sold by a firm named Biorepublik, who operated out of Tel Aviv, Israel. Previously, legal highs were regarded as a rip off, a hotch-potch of allegedly psychoactive herbs, caffeine and piperazines, the kind of products sold as Herbal Ecstasy at dayglo stands in festivals to the young, the gullible, or those facing drugs tests at work. Nobody knew what was in the Biorepublik products, but they ate and snorted them by their thousands. Australia, like the UK, geographically isolated and without land borders to any other nation, making smuggling harder, suffered the worst of the MDMA drought, as did the UK, and after the MDMA bust that year, the market there was starved. Cocaine is also extremely expensive in Australia – costing AUS$300, or around £195 per gram compared to £50 per gram in the UK – which is why the synthetic stimulant drugs market in Australia is so much more developed and entrenched than it is in the UK. Crystal meth-amphetamine use has grown there more dramatically than it has in Europe – three per cent of Australians over the age of four-teen have used the drug.

Biorepublik, run by a mysterious and elusive character, pseudonymously known as Doron Sabag, sold these new products as health supplements, in plain capsules. Sabag refused to reveal what was in them, setting a disturbing pattern that would be seen repeatedly in the coming years. Producers wanted to keep the formula secret to preserve their profits. And users were unanimous in their verdict: the NeoDoves and SubCoca worked, and they worked well. They made users feel amazing, briefly, and, at four pounds each, represented better value than the contaminated Ecstasy pills dominating the market. Customs officials were powerless to stop their import since they were sold as health supplements and their contents were not listed. What's more, they were being imported from Israel, not a country known for clandestine designer-drug manufacture and export, so packages were unlikely to be profiled and opened in any case.

The capsules sold in the hundreds of thousands worldwide throughout 2007 and 2008, and supply soon outstripped demand. Users were rhapsodizing about their MDMA-like qualities. That their sweat after taking them smelled like an old kipper left out in the sun was no great hardship. The smell was caused by poor syntheses and by the fact that the compound was also badly dried, meaning solvents and by-products, some of which smelt fishy, remained in the end product – so much so that Swedish users referred to the drug as *krebbe*, or crab.

The smell, along with the jittery comedowns, and the pounding heart and split lips and bleeding gums and ground-down teeth, didn't seem to much bother users. These drugs, whatever they were, were so much better than almost all the available illegal drugs available in the UK, Australia and Europe at that time. Someone somewhere was getting very rich, very quickly. Unsurprisingly, NeoDoves and SubCoca were also psychologically addictive. Some users reported going on unintentional binges for days, losing their minds and willpower, destroying their nostrils by opening the capsules and snorting them, 'fiending' through their whole supply – intending to have a single capsule and eating a dozen – chewing through their lips, twitchy and delusional, chasing the fleeting high. Then they ordered some more.

Their contents may have been a mystery, but it was a mystery worth solving for users concerned about their health effects after months of blithely swallowing the capsules, and for keen-eyed entrepreneurs who saw the huge profits available. A few months later, one of the compounds was identified on Bluelight as 4-methylmethcathinone, an utterly unknown chemical never before seen on the mainstream global drugs market. A poster at the Bluelight forum, phase_dancer, an Australian chemist working in harm reduction services for drug users was, along with others, responsible for the discovery. 'We've never been an organization as such, more a bunch of interested scientists working in different, but related fields. At the time, I was working as a chemist manufacturing reagent kits for Enlighten Harm Reduction, as well as researching improved formulations to better detect things like ketamine and PMA [a deadly impurity in MDMA pills],' he told me. 'No one seemed to know what the active ingredients were in the Biorepublik products. As popularity continued to grow, we put out word that there was a possibility drplatypus [a fellow poster] could arrange to have the products analysed. Samples were anonymously delivered to the hospital where he was working, which in turn passed them on to a colleague of ours from Adelaide Forensics.'

In the end several compounds were identified in the capsules, but the most important was 4-methylmethcathinone – soon to become known as mephedrone. Longer-term observers of the chemical underground, though, recognized this formula as the same one that Kinetic had devised in his home laboratory and posted to The Hive one bored evening back in 2003. The research chemical scene had gone overground – and how.

Mephedrone is chemically related to khat, or *Catha edulis*, a plant used for thousands of years in Arabic cultures, especially in Yemen and Somalia, as a social lubricant enjoyed for its stimulating qualities when chewed in a quid held in the cheek. Many shops in east London, home to immigrants from khat-using countries, sell the plant, which is legal and imported by established firms. It is brought in daily by air freight as it loses potency when less than perfectly fresh. The active ingredient, cathinone, is, if isolated and sold as a pure compound, a banned substance in most of Europe and is a Class C drug in the UK. Khat is banned in the US and some European countries, such as Holland, and its legality in the UK seems anomalous. It is perhaps overlooked because the number of people using it here is so small – around one-third of the UK's 100,000-strong Somalian community. In October 2012, 100 Somalian demonstrators petitioned Downing Street to ban the plant, saying addiction to the drug was causing family breakdowns and health problems.

Around 2004, in small kerbside newspaper kiosks in Tel Aviv and in nightclubs across Israel, a legal high known as *hagigat*, mixing the Hebrew words *hagiga*, meaning celebration, and *gat*, meaning khat, was gaining popularity. Hagigat capsules contained cathinone and were sold, legally, to buzz-hungry Israelis throughout the city. Cathinone, a chemical related to amphetamine, floods the brains of users with dopamine, the neurotransmitter that regulates mood, and which can be stimulated by safer methods including sex, good conversation and shared food – or by cocaine or amphetamines. Hagigat was banned in Israel in 2004, but continued to be sold more or less openly there.

In 2007 an unknown chemist associated with the Israeli firm Biorepublik had taken advantage of Israel's drug laws and, to beat the ban on cathinone, simply created 4-methylmethcathinone, a legal analogue of yet another cathinone derivative, methcathinone, and had sold the resulting drug as NeoDoves and SubCoca. Since the Hive was the most renowned underground synthesis board, and this synthesis and drug had never been seen on any recreational drugs market, it is overwhelmingly likely that the recipe was taken from the 2003 Kinetic posting on the Hive.

Granted, the chemistry is not complex: methcathinone is a derivative of methamphetamine, and 4-methylmethcathinone – mephedrone – is just the parent molecule with a few extra oxygens bolted on. For a chemist, the difference is perhaps a few hours' work, and analogous to a cakemaker simply decorating their product with a chocolate topping and a cherry. Whatever the truth of the drug's provenance in its 2008 iteration, most countries' drug laws were as inflexible as they are elderly and simply could not keep pace with these nimble online dealers.

Within a few months of the massive Cambodian oil burn, in February 2009, I wrote the world's first press reports about the MDMA drought, and the appearance of mephedrone and other new drugs in the specialist magazine, *Druglink*. My story was picked up by broadsheet and tabloid newspapers in the UK, who immediately gave the drug major coverage, catapulting it from an underground web cult to the news pages. Then things got even more bizarre, thanks again to the web.

With the formula out and published widely online, British dealers had found out what was in the Biorepublik capsules and had started to order the drug directly from Chinese factories, and set up websites to sell it.

Google's AdWords programme automatically generates advertisements from keywords paid for by subscribing businesses and places them on websites where those keywords appear, and who subscribe to its AdSense package. Some of these include newspapers. AdSense started helpfully adding links to mephedrone webshops at the end of serious newspaper articles, thanks to AdWords subscriptions by mephedrone dealers. Links to mephedrone telephone delivery services, with dealers on motorbikes offering to drop the drugs off at your home or office within a few hours, were generated automatically and published on the websites of newspapers such as the *Daily Telegraph* and the *Guardian*, which were demanding swift and decisive government action on the new killer drug menace. Mephedrone and other research chemicals received a vast, free boost.

Google does not manually or pre-emptively check what its advertisers actually sell before it accepts the advertisements, and the newspapers had no control of the ads that Google was generating for their pages.

While mephedrone was legal, Google also carried dozens of adverts paid for by the dealers on its own search pages. I approached the company and asked why, even at the height of the mephedrone story, adverts were appearing on its search pages. A spokesman replied, 'Under our drugs policy we do not allow ads for mephedrone. If we discover that ads are showing that break our policies, we will remove these as soon as possible.'

That afternoon, it took down dozens of advertisements, but refused to tell me how much it had earned by publishing links to sites selling the drug.

With this inadvertently brilliant marketing campaign by early retailers of the drug, dozens more websites sprang up selling mephedrone, many labelling it as 'plant food' in a bid to avoid medicine, drug and food labelling laws. Mephedrone took hold of Britain that year much as the drug hits users – fast and hard. There has never been a drug craze like it before or since, in chemical, legal, social and technological terms – but it's only a matter of time and molecular manipulation until another drug just like it appears.

'I prefer mephedrone to MDMA,' Dave Timms, a twenty-seven-year-old Londoner working in the fashion industry told me at the height of the craze in 2009. He was a sensible, intelligent, articulate guy who kept in good physical shape, dressed well, and liked to spend his weekends, from the second he left his office, getting as wasted as you can possibly imagine. He'd always liked taking drugs and partying, he said, but with the low quality of drugs available in the UK, mephedrone was just a better option – and far easier to get hold of. 'Mephedrone is more reliable, cheaper and actually more convenient than going to a dealer,' he said. 'I pretty much stopped buying coke and pills and MDMA once I found meph. I'd just make a bulk order and send off the payment and the package would arrive a few days later. I've been doing it for fourteen months, and have not experienced any negative effects, except sometimes I'm a little less motivated in work and training at the gym. It makes me laugh when I see people try it for the first time. Many are sceptical that something that's legal can actually work. But it does.'

He said he preferred mephedrone to MDMA as it gave him less of an emotionally fraught comedown after a weekend's use of the drug, and because it gave him greater mental sharpness. Timms once took it from Friday night until Monday morning without sleeping, and when he needed to straighten up and go to work, he simply snorted another line and put his suit on. He claimed during that honeymoon period with the drug that it had no downside, no comedown or hangover to speak of. And at ten pounds a gram, it was far cheaper than cocaine, which was selling then as now at around fifty pounds a gram for badly adulterated product.

Mephedrone completely wrong-footed the police, politicians, health workers and newspaper editors that year; they had little idea what the drug was or where it came from. It was the first drug that worked at the pace and in the manner of a web viral, following the now-classical narrative arc of digital marketing, with delighted users recommending the drug to their friends, who recommended it to their friends. The drug gained mass-market popularity in a matter of weeks, especially in small towns, where supplies of illegal drugs were more heavily cut or prohibitively expensive, while mephedrone was cheap, plentiful, uncut and 100 per cent legal. By contrast, MDMA took years to become so widely accepted and used.

Many sites that sold it referenced rave culture in both their graphic design and imagery, and it was clearly intended for use as a narcotic. Sites such as Ravegardener, Champagnelegals, Bubbleluv and market leader UKLegals sold the drugs by the vanload. Items to be sold for human consumption must pass food safety and medicine laws, but as with the earlier research chemicals, as long as mephedrone carried labels saying it was not to be consumed, vendors could avoid any of the customer protection laws that the UK had in place, as well as the Misuse of Drugs Act, which did not cover this previously unseen, unheard of beta-keto methcathinone.

A kilo could be imported to the UK in a matter of days from China, and customs officials were powerless to intervene. Now that there was serious money to be made, more professional outfits stepped into the market, and started selling it blatantly on public websites, with slick design, smart-ordering systems, sharp back-end databases and overnight delivery, one-click orders and thousands of positive online reviews.

Soon dozens of Chinese labs were pumping out tonnes of the drug every week, feeding a voracious demand. It was a move of considerable ingenuity on the part of many British manu-facturers to avoid setting up laboratories here in the UK with all the attendant risk and expense, and simply to outsource production to a country where the well-established chemical and pharmaceutical industry was willing to turn a blind eye to sketchy export dockets, where local officials could be paid off cheaply, and where labour costs were lower. It had worked in every other manufacturing sector so why not for designer drugs, too? Legitimate Chinese exports to the EU in 2003 were 106.2 billion euros. They more than doubled to 231 billion euros in 2007. By 2010 they stood at 292 billion euros – a threefold increase in under seven years. Business, both legal and para-legal, was booming.

The owner of one factory sent me a list of his consignments to the UK at the height of the mephedrone phenomenon in February 2009. Posing as a bulk buyer for an undercover report for the *Mail on Sunday*'s *'Live!'* magazine's Reportage slot, I had demanded references from satisfied clients. He was more than happy to help. He revealed that he had sent fifty kilos of mephedrone from Shanghai to the UK in a single week, by

Fedex and DHL, through Charles de Gaulle airport via Gatwick and Stansted, to every corner of Britain, and hundreds more kilos all across Europe. As the UK shivered under its first real snowfall in a decade, the country was buried under an avalanche of a very different white powder. Within a few months, Chinese mephedrone had Britain and Europe in its fierce, eye-rolling grip; it was like the Opium Wars in reverse.

The 2011 *MixMag* drugs survey, the world's biggest, clearly showed the explosive growth of mephedrone. The majority of the more than 2,000 drug users recording their habits for the previous year for the clubbers' magazine said they had taken the drug, which now ranked fourth behind Ecstasy, cocaine and marijuana. There had not been a new drug since Ecstasy in 1988 that had such immediate and dramatic effects on patterns of consumption.

That year, other modified drugs also came on to the market. There were new analogues of mephedrone itself, such as methylone, butylone and pentylone, and other stimulants, such as the powerful MDPV (active at just a few milligrams and responsible for users posting reams of alliterative nonsense online); there were flephedrone and buphedrone, brephedrone and dozens of other drugs in the cathinone family: all emerged in rapid, dizzying succession, all potently psychoactive, and all completely legal. Research chemicals had now been re-marketed, with considerable skill, as legal highs. Some websites sold the pure chemical compounds, others sold branded sachets that carried no information about their contents.

Marijuana replacements also came into vogue between 2008 and 2010, as alternatives to a popular, but illegal drug. Again, the research chemical scene and Chinese laboratories were responsible for the appearance of these new, untested drugs. Before that time, anyone who bought legal alternatives to marijuana was almost guaranteed to have been sold an inert substance that would irritate the lungs, empty the wallet and do very little else. Given the easy access to marijuana in most of the world, the market for such function-free products was narrow, limited solely to the very young, the gullible, the exceedingly stupid, or the very cautious.

But at more or less the same time as mephedrone appeared, reports started to emerge on dozens of drug forums that a new synthetic marijuana product, named Spice, was actually very powerful, and that it smoked very much like marijuana. News of its potency spread around the web as quickly and pungently as billowing gales of ganja smoke through a festival crowd. But just as with mephedrone's first appearance in the NeoDove capsules, nobody knew what the active ingredients were in these bags of herbs.

The bags of Spice were sold for about fifteen pounds each for the classic pot dealer's measurement of three and a half grams, or one-eighth of an ounce. Manufacturers claimed a hitherto unknown synergy between the ingredients, which were listed as baybean, blue Egyptian water lily, skullcap, lion's tail and sacred lotus, albeit in the original Latin for added authenticity and confusion. The real truth of Spice's power lay in the laboratory of a brilliant chemist named John William Huffman.

Huffman is now eighty years old, and has recently retired after a long and distinguished career as Professor Emeritus in organic chemistry at Clemson University in South Carolina. You'd never guess that this elderly gentleman, with his tidy beard, plain spectacles and owlish manner, is responsible for getting thousands of people incapably stoned. On a mild spring afternoon in 2012, Huffman was kind enough to speak to me while relaxing after a recent bout of painful surgery. The professor chuckled down the Skype line mellifluously, sometimes gazing at the nearby Smokey Mountains, as I asked him how the Spice story happened.

Between 1984 and 2011, Huffman and his colleagues had created over 400 synthetic cannabinoid compounds while studying the structure-activity relationship between a series of compounds that resembled tetrahydrocannabinol, the active constituent of marijuana, and the human brain. The human brain has cannabinoid receptors, and the molecules that are found in marijuana and hashish, such as THC (a highly active constituent of the drugs), act as keys to open those locks. Huffman wasn't looking to create a psychoactive compound in his research – quite the opposite. He was following in the research footsteps of American firm Sterling Winthrop Pharmaceuticals (SWP), who were trying to develop non-steroidal anti-inflammatories, a class of drugs to which aspirin belongs. One of the compounds SWP discovered turned out to be a weakly active cannabinoid. Huffman's team's work was based on this series of chemicals, and the new compounds they produced were tested on rat brains. 'We were trying to relate chemical structure to biological activity,' says Huffman. 'The way you do this is to make a series of compounds, varying the structure from one compound to the next, and we were then trying to figure out how these interact with the same receptors that THC interacts with.'

His motivation? 'Pure science, scientific curiosity,' says Huffman. 'As it turns out, the human endocannabinoid system, as it's called, has profound effects on human behaviour, pain, mood, nausea and appetite, and lots of other important biological functions. And here we have these compounds that don't look anything like THC, and they don't look like the endogenous cannabinoids, but they did have activity.'

In 2008 a German newspaper sent the mysteriously effective bags of Spice for nuclear magnetic resonance analysis, which peered into the molecular structure of the sample and definitively identified it. Spice did not work through any herbal synergy; the active component was JWH-018.[4]

Somewhere, somehow, this experimental medicine had escaped from the medical journals where it was published in 2006 and was being sold for profit on the research chemicals market. More brands quickly appeared, such as Black Mamba, SKUNK! K2 and Abama. Clumsy though these brand names may be, they're certainly snappier than (1-pentyl-3-(1-naphthoyl)indole), or JWH-018. The compounds were soon being exported from China in massive bulk to the US. Thanks to the relentless hedonistic imperative, and laws prohibiting the use of marijuana in much of the world, in the space of a couple of years these herbal mixes sprayed with JWH-series drugs would be seized by police in every city in the US.

'My immediate reaction [upon hearing people were smoking these compounds] was that I thought it was humorous,' Huffman said. 'I heard somewhat later that these compounds have extremely bad side effects and they are much, much worse than marijuana. The synthetic compounds seem to cause some serious psychoses.'

Some dealers then dropped the herbs and started synthesizing the pure compound and selling it by the gram. As the compounds were legal in the US and UK and Europe, Chinese labs would send kilos of the white powder under plain cover for a few hundred dollars. Profits were around US$10,000 to US$20,000 per kilo. The synthesis of the compounds is relatively simple, and with a digital grapevine trembling loudly with the news of 'legal pot', thousands of sites appeared in a matter of weeks. Some of them sold what they promised, others were rip-offs, but all of them made large amounts of money by selling an array of Huffman's compounds – even some of the inactive ones. A very early JWH-018 pioneer fills in the gaps. 'It started out with just me, then a friend and his partner helped out – they'd been in the RC [research chemical] business since the early to mid-nineties. They gave me access to resources I would never have had otherwise, such as the ability to produce and ship JWH-018 in the USA on a monumental scale.'

Setting the company up wasn't without its trials, he says, and an early problem was payment and banking. 'PayPal never allowed research chemicals traders, and they're bastards with your funds if you get caught by them. I managed to trick them for nearly two years by developing a fairly convincing yet totally fake website for JWH-018 "Bonsai fertilizer", which was guaranteed to produce bigger and taller bonsais. Yeah, I was the original "Bonsai food" vendor!'

Online today, young American users persist in discussing how they have 'applied the material to their plant's lungs with great success', believing that this cunning subterfuge will outwit any jury and judge thrown at them.

But perhaps before mocking them, we might consider that drug laws in the US are among the most punitive on the planet. America's incarceration rate is the world's highest, and that shameful statistic has been largely driven by its war on drugs. America's lack of credible nationwide action on drug law reform, in particular around decriminalizing marijuana, has turned many thousands of its citizens into unwitting lab rats, self-administering chemicals more dangerous and untested than the compounds they are substitutes for. In 2011 many of the new cannabinoids were banned in the US on a state, but not federal level. And again, the scientists simply went back to the labs, or the web, or medical literature, and cross-referenced the new compounds with their local drug laws.

Likewise, in the UK, as soon as the Misuse of Drugs Act had been amended to include the new cannabis-like substances a year earlier, in 2010, many more appeared. One of the new drugs that replaced JWH-018 in the UK was AM-2201, a particularly powerful compound active at such a tiny level – just a milligram – that overdoses were virtually guaranteed. This would not have been the intention of its creator, the eminent biochemist Alexandros Makriyannis of Northeastern University, Boston, whose work investigating the body's cannabinoid receptor system is as respected as Huffman's.

On 1 March 2011, the DEA temporarily placed five synthetic cannabinoids (JWH-018, JWH-073, JWH-200, CP-47,497 and

CP-47,497 homologue) into Schedule I in the US for one year (extendable by six months). But hundreds, if not thousands of possible analogues of Huffman's and Makriyannis's work remain legal, their effects completely unresearched. And there are many Chinese laboratories willing to send these simple compounds to the West, making thousands of dollars in the process.

A poster named Where Wolf told the Bluelight drugs forum in 2008 why he smoked Spice instead of marijuana.[5] During one of the unpredictable yet regular marijuana droughts that seize even the world's largest capital cities, he had, he said, got fed up of smoking adulterated grass. In 2008, marijuana dealers in Holland, and Vietnamese growing gangs, particularly in London but also in the rest of the UK, had discovered that by spraying small silicone beads onto freshly harvested plants they could increase weight – and profits – by up to 20 per cent. What's more, the small glassy balls looked like crystals of THC to the naked eye, their glistening globules promising a strong smoke. This adulterated grass became known as 'grit weed'. It was dangerous to burn and inhale, but that didn't concern the dealers much, since their profits were boosted by the extra weight. 'I'm based in London at the moment, and have pretty much quit weed for Spice entirely now,' wrote Where Wolf. 'This is partly because, though I have a range of sources all over the city, quality has dropped significantly in the last few years: for all the press hysteria about killer skunk, I damn sure haven't had any out-standing weed in years. Previously great sources have become so-so.'

He also said the drug laws banning marijuana had driven him online in search of an alternative. He wrote:

I hate carrying [marijuana] on public transport: I'm part Middle-Eastern (Israel: Arab-Jew), and get searched quite a lot. Random use of sniffer dogs at Tube stations is pretty common these days. I discovered Spice when the market first went to hell, and was amazed to find it worked. Tolerance does build quickly, but when you haven't smoked anything for a while, 2–3 spliffs can produce a real glowing body-high. I'd say it's as pleasant, if not always as potent, as a lot of the hybrid pseudo-skunk I've smoked over the last five years. Seems a really sad comment on the UK cannabis scene that there's a legal product that's almost as good.

It may have been as effective, in certain ways, but it certainly wasn't as safe as marijuana, says Huffman. 'The synthetics are much more dangerous,' he explained. 'No one has ever died of an overdose of marijuana. You'd probably forget where you put the stuff, because it has an effect on memory. These synthetic compounds, they interact differently with the cannabis receptors to marijuana. They have the same effects at a superficial level, but marijuana and THC lower blood pressure, whereas some of these compounds raise blood pressure dramatically, but we don't know why.'

In the US there exists now the most knottily tangled of legal situations. Keeping marijuana illegal means people are taking largely unknown compounds of unknown strength with zero toxicology reports even in animal tissue, much less actual animals, since smoking them is less likely to end in a life-ruining jail term. Still, in the US, where US$23 billion is earmarked for the drug war in 2012, there's a lot more money than logic going around. And it can only be described as a crisis of integrity that while President Obama has spoken openly of his habitual and much-enjoyed marijuana use in his youth, his administration offers no logical, scientific basis for the retention of the ban.

By the middle of 2009, the mephedrone market had turned into an even more frenzied free-for-all, with new vendors popping up daily. Quick profits were guaranteed, as mephedrone and other analogues could be bought for around £1,000–£2,500 per kilogram in China, and sold for £10,000 perfectly legally. Fedex, USPS, UPS and other international couriers were soon unknowingly sending multi-kilo consignments of designer drugs all over the UK and Europe to retailers. 'We can send this under plain cover, marked as deodorizing crystals for babies' nappies to you in Britain,' said Eric, a major vendor who operated out of Shanghai.

Next-day delivery to individuals via the Royal Mail turned thousands of postmen nationwide into unwitting drugs couriers. Facebook groups dedicated to the drug sprang up, and pubs and clubs all over the UK began to reek fishily of mephedrone, which was sweated out as gurning users danced and ranted. When a Facebook page called THE MEPHEDRONE EXPERIENCE! appeared – presumably set up by an excitable user under the influence – it quickly gathered dozens of members, too stupid, high or careless to realize that they were sharing information about their drug habits with their workmates, friends and family. Many users were still so unaware of the drug's exact effects or its power that they could be seen taken by surprise in the least likely of places. Even an upmarket private members' club in London's West End had a few staff giggling, saucer-eyed, over-tactile and unable to add up a two-drink order correctly, on the night I visited in early 2010.

Many internet forums dedicated to selling and discussing mephedrone and other new legal highs came online, with hund-reds of thousands of page views in a month. An early gathering point for mephedrone users was Champlegals.co.uk, a forum attached to the website by the same name. Threads there stretched out to a mindboggling size, with tens of thousands of page views. 'Juice soldier', posting on the Bluelight drugs discussion board, spoke for many users when he said on 4 April 2007 of the Biore-publik range, 'After trying them properly, I can honestly say I got more of an md[ma] like buzz out of these than any pill I've eaten in over 12 months. That's got to be saying something. Doesn't last as long – but they are easily as effective if not more, not [heavy] like MDE – clean and euphoric like real MDMA.'

No one had the vaguest idea what the long-term effects of mephedrone and all the other new drugs were. People couldn't even agree on what to call mephedrone: as its chemical name could be rendered 'M-MCAT', users on Champlegals jokingly said it could be called 'Meow'. Soon enough, that information turned up on Wikipedia, was quoted verbatim by lazy journalists on short deadlines, and entered the popular culture rapidly. Until the newspapers called it Meow, no user ever did, seriously.

A simple web analytics exercise using Google's tools showed an Everest-like peak for users entering the term 'mephedrone' into the search engine within weeks of my first stories appearing. But more tellingly, the tool showed an even more pronounced spike in the use of the term 'buy mephedrone'. Later research would show that searches motivated by the desire to purchase the drug actually peaked most dramatically following news reports of alleged deaths from the drugs.

Like all parties, the mephedrone craze had to end, or at least wind down to the last few die-hards. Disturbing stories started to emerge. Users reported that their knees and fingers had gone purple, whether because the drug caused severe cardiotoxicity and vasoconstriction (narrowing of the arteries), in common with other cathinones or amphetamines, or because they were paranoid. Hospitals also reported a sharp increase in admissions of users suffering heart palpitations. Dr Adam Winstock, consultant addictions psychiatrist at the South London and Maudsley NHS Trust and honorary senior lecturer at the Institute of Psychiatry, King's College, says the mephedrone honeymoon was short. He told me, 'Two years after first encountering mephedrone and considering it as a relatively benign substance, I now think it has a deeply unpleasant harms profile with a high risk of abuse and dependence in many users. Users rank it as more unpleasant and risky than either cocaine or MDMA. It is early days still and there is lots we don't know – but it is not a safe alternative to MDMA.'

In December 2008, an eighteen-year-old Swedish woman in Stockholm had become the first woman in the world to die after using the drug. A day after her death, the drug was banned in Sweden. The first media reports of deaths in the UK appeared in November 2009, when Gabrielle Price, a fourteen-year-old from Worthing, in West Sussex, became ill at a house party where she had taken the drug together with ketamine. At first, it was claimed and widely reported that she had died as a result of the drug. However, a pathologist's report showed the girl died of broncho-pneumonia following a streptococcal A infection. This kind of inaccurate reporting was repeated with grim regularity. By July 2010, fifty-two deaths were claimed to be associated with the drug in the UK, but coroners' reports eventually and conclusively showed that over a quarter of the supposed victims hadn't taken the drug at all, and no clear data existed to prove that mephedrone caused the deaths of those who had. Fiona Measham, a member of the ACMD and a senior lecturer in criminology at Lancaster University, says the reporting of the unconfirmed deaths by newspapers followed the usual cycle of 'exaggeration, distortion, inaccuracy and sensationalism'. Not since the death in 1995 of fifteen-year-old Leah Betts (the Essex schoolgirl who died because she drank many pints of water after taking Ecstasy, fearing she was dehydrating) had the media launched such a concerted, and misinformed, campaign around a new drug.

There is no doubt that the drug could be dangerous, as Winstock and other doctors have asserted with insights gleaned from two years of working with users, but media reports at the time were so misinformed and badly verified that they often had an air of satire. Some tales were laughably implausible, yet they were repeated verbatim, becoming part of the folklore. Newspapers reported how one user in County Durham in November 2009 had torn off his own scrotum after hallucinating for eighteen hours on the drug. The paper failed to mention that the press release on which it was based actually gave the source of that claim as a spoof testimonials section of a website selling the drug.

The media reports of deaths continued, and a tipping point was reached in March 2010, when two young men from Winteringham and Scunthorpe, eighteen-year-old Louis Wainwright, and nineteen-year-old Nicholas Smith, were found dead at their homes. Later investigations found the pair had actually taken methadone, the potent heroin substitute that can slow breathing dramatically in users with no tolerance to the drug, leading to death. It is unknown whether Wainwright and Smith believed they were taking mephedrone, but certainly many people mistook the word 'mephedrone' for methadone around that time. No matter. Politicians had to be seen to act fast, and in the charged atmosphere of a tabloid feeding frenzy, it's hard, on reflection, to blame them completely.

The ACMD had a torrid year in 2009. Chairman David Nutt was sacked in October following his statistically accurate observation that Ecstasy was less likely to kill users than horse riding, and after his call for a new discussion on the classification of cannabis was rejected by the Home Office. There were five resignations in support of Nutt's views. In March 2010, another member, Dr Polly Taylor, resigned in protest when the Code of Practice for Scientific Advisory Committees – new government guidelines, widely seen as impinging on scientists' objectivity – were published. Science, the government had decided, was now the servant of policy. The ACMD had been weakened by these events and when the debate and press panic over mephedrone began, it was already on the back foot.

In April 2010, after months of increasing media hysteria and misinformation on drugs generally, and mephedrone in particular, the government felt compelled to legislate. Usually, the ACMD would be allowed time to investigate and research a drug's harmfulness and then offer its expert guidance to politicians. But the ACMD was pressured by the government to announce that the drug was harmful and should be made illegal. The British government then banned all the substituted cathinones, the chemical family to which mephedrone belonged. ACMD member Eric Carlin, an executive consultant in the drug and charity sector, immediately gave up his seat in protest, and said in his resignation letter:

> We had little or no discussion about how our recommendation to classify this drug would be likely to impact on young people's behaviour. Our decision was unduly based on media and political pressure ... I am not prepared to continue to be part of a body which, as its main activity, works to facilitate the potential criminalization of increasing numbers of young people.[6]

In the April 2010 edition of British medical journal *The Lancet*, a leader writer lamented the 'collapse in integrity of scientific advice in the UK', berating the government for political interference in its haste to ban the drug without proper expert consultation. 'The terms of engagement between ministers and expert advisers endorsed by Alan Johnson have been blown apart,' wrote *The Lancet*. 'During the past 12 years the Labour government has done a great deal to build up a strong science base in the UK and enhance the important role that science plays in our economy and society. However, the events surrounding the ACMD signal a disappointing finale to the government's relationship with science. Politics has been allowed to contaminate scientific processes and the advice that underpins policy,' it said.[7]

China bowed to pressure from British institutions, including

SOCA, and banned the manufacture, export and possession of the drug in August 2010. Mephedrone had now been banned in Austria, Belgium, Denmark, Estonia, France, Germany, Ireland, Italy, Latvia, Luxembourg, Malta, Poland, Romania, Sweden and the UK. The European Commission advised the remaining EU countries to ban the drug in October that year. 'It is good to see that EU governments are prepared to take swift action to ban this dangerous drug,' said EU Justice Commissioner Viviane Reding after the twenty-seven-nation bloc agreed the total ban, imposed in December 2010.

John Ramsey of St George's, University of London, is an advisor to the ACMD. He also runs a commercial organization, TICTAC Communications, which produces and sells a database of chemicals bought from the internet or retrieved from amnesty bins at raves, festivals and clubs across the UK. Ramsey tests the drugs, and then publishes the database on a CD which police forces and hospitals use to identify the torrent of new tablets, powders and capsules that are washing up across the UK and Europe. He was among the first to identify mephedrone in UK pills and powders bought on- and offline. 'Do come and look at the archives – did you notice on your way in?' he said to me as I visited his laboratories at the hospital in May 2012. He showed me a huge row of metal drawers. Inside each, labelled with the careful exactitude of a taxonomist, were samples of drugs that this lab has tested. There were 27,000 samples of different drugs here, gathered by this careful and dedicated man. With his bookish air, smart dress and a keen and scholarly attitude, he was the least likely procurer of drugs I had ever met.

'Ah, yes,' Ramsey mused over one small vial. '2C-B. There's about 400 quid's-worth here. I do wonder why they threw them away. I wonder what story lies behind these?' he said, looking at the small butterfly-stamped tablets that were seized from an amnesty bin at a rave. 'Why would they throw away such a huge amount of drugs? It's baffling.' Ramsey has the most comprehensive stash of drugs on the planet – and it's very likely that he'll soon need new premises. Since the discovery of mephedrone, the situation involving new drugs is ever more complex, and evolving faster than anyone can keep track of, he says. 'It's drug control that spawns it all to some extent. The link to the whole thing is the Chinese chemical industry, the ability to scale up from what you can do in a bedroom set-up to an industrial level,' he says. 'I'm sure there are people who, like Shulgin, tinker with stuff and then take it, but it's never going to become mainstream unless you can make it in reasonable quantities.'

If high-level pressure were to be placed on the Chinese, he says, it wouldn't be long until production shifted somewhere else. 'It would inconvenience them slightly, but it'd only be a matter of time until they moved elsewhere.' One vendor supported Ramsey's view, when he told me he believed the Chinese authorities were complicit in the mephedrone trade, until it started to gain Chinese users. 'It all contributes to their surging economy,' he said. 'In the case of mephedrone, there was some external pressure from the UK government, but also, it was starting to show up in clubs in Taiwan, Hong Kong and even China. That's probably why it was actually banned there. We tried to get a lab going again underground, but the Chinese authorities were right on top of it, even after bribes were offered, and so we had to give up,' he says.

Ramsey's laboratories could not be further removed from the muddy fields and laser-lit clubs where these drugs were seized or surrendered. Bags of pills and white powders marked 'V Festival, 2011' and 'Glastonbury 2011' lie on the workbenches. There are bags with online vendor names from all over the world, including Taiwan and the US. 'We sometimes buy in euro or pounds and they get sent in to us from Belize, all sorts of places,' says Ramsey. TICTAC's most recent research reveals much about current patterns of drug use in the UK, especially for those who feel this is a small market.

Ramsey and his team asked anonymous volunteers to use a temporary toilet they had placed in the car park of a major London club. Analysing the urine afterwards, he found there were over thirty-five different metabolites in the waste, including many of the new drugs mentioned in this book. 'It's very widespread,' he says. 'The data is complex, but if I can just stick a toe in the water here, the problem is very widespread – and growing.' A half-beat pause, then he cracks up in laughter. 'Poor metaphor.'

To identify the chemicals in the powders and pills, Ramsey has them crushed to a fine powder, then pinned with a diamond against a transparent plate, illuminated from below with a beam of infrared light. The resulting spectrum of each compound has a distinct signature, a translucent splash of psychedelia, which is then catalogued; check the library and if the image matches, it's a drug. Gas chromatography is the next test, which splits the compound in a solvent and then analyses the gases. Then it undergoes mass spectrometry, during which a beam of electrons is smashed into the molecules, which are then ionized. That data is parsed electromagnetically into the substance's likely chemical composition. Finally, if it's still not clear what the chemical is, it's sent off to the nuclear magnetic resonance machine, which 'sees' into the molecular structure and identifies, once and for all, what it is.

Under the strip lights, there are also thin slices of rats' brains kept alive in buffers fed with oxygen. Here, the new drugs that land in the laboratories have their first ever empirical and formal testing. This is pharmacokinetics, the study of the drug and how it reacts in vitro. Rats' aortas, cut from their hearts, are flooded with the chemicals Ramsey and thousands of users worldwide have bought online. There are serotonin receptors in mammalian hearts, and these and the rats' brain slices are monitored by micro electrodes that measure the amount of serotonin or dopamine released, as well as the pharmacokinetics – how the released neurotransmitters move around the tissue.

This, other than taking it yourself in a Shulgin-esque game of chemical Russian roulette, gradually increasing from a small dose to a larger dose over a number of days, is the only way to tell the active dose of an unknown chemical. And this, perhaps, is the key issue in drug use today for many people. They are acting as Shulgin did – testing unknown compounds, but with the difference that they neither made the drug nor have any real idea where it came from. Receptor binding studies such as those carried out by professional pharmaceutical laboratories are vital, but they are a complex task of the kind unlikely to be commissioned by either Chinese vendors or resellers on the net, even though their cost – at between £1,500 and £2,500 per sample – is a fraction of the profits made by the illegal laboratories that come up with the new drugs. But illegal drug salesmen tend not to be renowned for their altruism.

The drugs are getting into the UK in serious bulk because the Border Agency is overworked and underfunded. Once here, they find a ready market. 'This is a trade like any other,' says Ramsey. 'You need several components: an innovator to think of structures, a manufacturer to produce it, marketing and advertising.' Those elements combined made the research chemical scene burst on to the high street in 2009, in a drug-starved country reeling a from a banking crisis brought on by a lack of regulation. The cathinones, with their speedy, cocaine-meets-MDMA-like buzz, were the perfect drug for those outlaw times. Mephedrone was credit crunch cocaine, neither one thing nor another, an analogue of an analogue. The Labour government, its drug policy no more radical than the previous Conservative administration's, swung into action and banned it, driving a coach and horses through both scientific concerns and all political protocol. Mephedrone is harmful if taken to excess, and many users could not control their use, but questions about its relative harmfulness were never properly asked of the scientists who are specifically charged with establishing those harms. That led, in the minds of many users, to a deepening mistrust of the government, whose politicized messages on drugs no longer existed in a vacuum of information, and were subverted by the views of users on the web.

The anti-hierarchical structure of the web, created in an act of deliberate subversion and anarchy by countercultural, drug-loving hippies, instead offered the curious a range of voices, of experiences, and value systems that laid bare the prohibitionists' stance as a flimsy charade. The dominant discourse was disrupted – permanently. Official responses over the coming years were often short-termist and morally panicked, but the bigger story was the change in consumer habits, and drug habits. Mephedrone changed attitudes to drugs overnight. People who had never heard of Shulgin, who would not previously consider buying a drug online in the original research chemical scene, who had never read a trip report much less likely written one, were taking research chemicals they had sourced on the web in the belief that they were safe because they were legal – or not caring either way.

The mephedrone ban had a series of unintended but wholly predictable consequences. Rather than reducing harm; rather than limiting the production of new replacements; rather than reducing demand, instead, it increased all of these. All the circles closed in 2009–2010. The web had become a space where many of our cultural and economic transactions were taking place. Social media, knowledge-sharing and ecommerce were also now the norm. Drug use in the EU was so commonplace as to be an epidemic. By the middle of 2010 as mephedrone was banned in the UK, the scene was set for the next stage in the Drugs 2.0 revolution. The search for mephedrone's replacement began in earnest, with users and dealers outdoing each other in ingenuity and greed.

# 7

## *Woof Woof Is the New Meow Meow*

After mephedrone was banned in the UK and the EU in 2010, supplies dried up as vendors closed shop to avoid prosecution. Many vendors and users searched for replacements, but the majority of the next wave of research chemicals – or, as they had now been cleverly rebranded, 'legal highs' – were dangerous, ineffective and poorly made. The already vast and varied offering of new drugs rapidly grew, and often contained unknown stimulant blends. Many users blindly consumed new drugs that made even Chinese mephedrone look safe by comparison. Like the users, the tabloid press in the UK, too, were desperately seeking a mephedrone replacement. Editors know that drug hysteria sells newspapers, both to worried parents and to desperate or curious users.

Newspaper editors, policy-makers, politicians, pundits and police had all failed to notice the strands linking the research chemical scene and the legal high market, the psychonaut to the high street. What had knitted the two together into such a tangled mess was mephedrone, and what had enabled that entanglement was the prevalence of social media and internet use among user groups and retailers. The research chemicals market was back – and this time the people involved were neither obsessives, nor evangelists, nor psychedelic philatelists collecting new drugs for fun. This time, it was all about the money.

Previously the research chemical market had been self-policing, and with a small number of vendors and users the risks were contained to a small group and generally, if not universally, understood. That's a view confirmed by a European research chemical vendor, André*: 'The scene has drastically changed. Back in 2003 it was unknown both in the street and the internet, there were just a few users spread here and there who knew about the books *PIHKAL* and *TIHKAL*,' he told me. '[The scene] was mainly around a few discussion boards where the admins ran their own shops in a private way. After the mephedrone hype, it became public everywhere, in the media, on the streets, social networks, YouTube. Now this scene is not underground any more, as it should be. Everyone tried to make fast money out of mephedrone, there were kids selling this chemical everywhere for a ridiculous price. The market was flooded . . . Of course these people were not even [aware of] what they sold.'

One branded powder, Ivory Wave, looked likely to become an early candidate for a mephedrone replacement. Originally, when mephedrone was legal and many new substituted cathinones were appearing on world markets for the first time, it had contained MDPV – a powerful cathinone-related stimulant active in tiny doses (around 5–10 mg) which, if taken in excess, causes psychosis.

The manufacturers did not list the ingredients on the label and the truth would only be found out later, when analysis had been carried out by professional organizations. The makers of this branded legal high also mixed in numbing agents that mimicked cocaine but had no psychoactive effect – a hugely irresponsible move, since typical lines of cocaine are around ten to twenty times as large as the active dose for MDPV, and presenting the drug in this fashion led users to take far too much.

---

* Not his real name.

When MDPV was banned along with mephedrone, the Ivory Wave manufacturers maintained the name of the brand, but switched the ingredients to include a different, legal compound, desoxypipradol, which belongs to yet another category of drugs, the even more potent piperidine class of stimulants. The drug is active at tiny doses of 1 to 2 mg and its effects last for twenty-four hours. Many people who had been buying mephedrone, now accustomed to buying legal highs online, bought Ivory Wave instead. Its sale coincided with a sharp increase in hospital admissions for people suffering from psychosis and paranoid delusions. Twenty-four-year-old chef Michael Bishton, of Ryde on the Isle of Wight, was the first victim, found dead in August 2010 after jumping to his death from a cliff top, reportedly suffering from paranoid delusions. He died before he learned that his fiancée was carrying their first child.

Dr Kate Willmer, consultant cardiologist at West Cumberland Hospital, told the *Daily Telegraph*, 'People are coming into the hospital in an extremely agitated state, with acute paranoid psychosis. If you try to give them anything to help them, they are convinced you are trying to harm them so we have had to completely knock out two or three of them in order to treat them . . . it is taking two to three days for the agitation and psychosis to wear off. I have never seen anything like it.'[1]

Mephedrone refugees were in for an exceedingly tough time in the coming years, their inexperience and their lack of knowledge and respect for these powerful chemicals almost guaranteeing horrifying experiences. A comprehensive pharmacopeia of the years following mephedrone would be several volumes long, but a few statistics give a sense of the scale of the issue.

In 2010, the Lisbon-based European Monitoring Centre for Drugs and Drug Addiction (EMCDDA), said in its annual report that in the thirteen years between 1997 and 2010, just 150 new drugs appeared on the market.[2] Record numbers of new drugs appeared in 2009, when twenty-four new compounds were discovered for sale in a single year. In 2010, there were forty-one. In 2011, 150 new drugs were initially reported, though later figures would revise this down to forty-nine. By early March 2012, there were already thirty-six more – a staggering 150 confirmed new compounds in just three years, with no indication that the market would slow down.

An early mephedrone replacement was, supposedly, naphyrone. It was branded as NRG-1 and vendors refused to reveal its chemical structure for fear of others copying the synthesis and capturing their market, as had happened with Biorepublik's NeoDoves and SubCoca. Its manufacturer promised vendors that this new drug had a substantially similar effect to mephedrone, and many of them bought it and sold it to hundreds of their customers. Unhappy reports surfaced almost immediately, saying the experience was unpleasant at best, and dangerous at worst. No wonder: a *British Medical Journal* piece in June 2010 found that of the twelve legal highs it analysed, most sold as NRG-1 or NRG-2 contained either a banned cathinone, caffeine or inactive ingredients.[3] Many online dealers were selling anything and calling it a legal high, or NRG-1, naphyrone or any other eye-catching name, not knowing or caring what they were selling. Users were just looking for a new drug to replace mephedrone.

In June 2010 the *Cambridge News* regurgitated uncritically the news from the website of a local headshop, Cambotanics, that it was selling a fantastic new legal high, this time, known as 'woof woof'.[4] It was in fact MDAI, a non-neurotoxic derivative of MDMA developed by chemist Professor Dave E. Nichols at Purdue University, which he made as part of his work studying serotonin receptors. But the effects were reported to be mild at best, and the drug did not catch on.

The laboratories in China which are producing these new drugs are normally legitimate companies that also produce generic drugs, such as anti-retrovirals for HIV and heart disease drugs, or that are dedicated to the production and sale of fine chemicals, raw materials, pharmaceutical intermediates and biological products. In both kinds of set-up, workers, normally young chemistry graduates, either carry out syntheses after hours, diverting the necessary precursors from their use in the production of legitimate products and keeping the money themselves, or they work with the complicity of the laboratory owners, who take a cut of the profits.

Payments are made to uninvolved third parties, often using global money transfer giant Western Union, and shipments are carried out by legitimate courier firms such as EMS, Fedex, TNT or DHL. Bribes are paid to Chinese officials to avoid the need for paperwork or export permits. Multi-kilo quantities are labelled falsely as nappy deodorizer or carpet cleaning chemicals, or, in the case of small orders, they are secreted in envelopes and marked as documents. These factories deal with individuals, or with retailers in the UK, the US and the EU, some of whom repackage the drugs they receive into branded legal highs, while others sell the compounds by the gram.

'Eric', a Shanghai lab technician, is a typical manager of such a facility. In a sting operation I set up for a report in Britain's *Mail on Sunday*'s '*Live!*' magazine in April 2010, journalist Simon Parry visited the laboratory that was, at the time, producing most of China – and Europe's – mephedrone.

After winning Eric's confidence over a number of weeks in a lengthy email correspondence, in which I posed as a large, London-based wholesaler of legal highs, I had made an order for 120 kg of MDPV – enough for 2.4 million doses. I chose MDPV deliberately, as it had been banned in the anti-cathinones legislation in the UK, and I wanted to see what effect the ban had had on the supply-side of the new drugs. Eric told us he was able to ship the products via express courier to a 'soft' country such as Greece, from where the products could be trans-shipped. 'We have agents in Europe so we can send to Ireland, Austria,

Spain and Italy. Then the package will be re-sent to the UK from those countries. If the package comes from outside Europe there might be trouble. Within Europe, the UK customs normally will not check,' Eric told Parry. 'If it is stopped we always refund or reship. That is why we have so many customers in the UK. There is no risk for them.'

I insisted on a site visit to the lab before confirming the order. Parry visited the factory as my local agent, keen to examine the facility, but in reality, it was to interview Eric in person and to take photographs of the inside of the laboratory that had, up until the ban in the UK, supplied hundreds of kilos of mephedrone a week.

Eric's laboratory is the size of a small flat, and is near the city's international airport, with neighbours including offices of GlaxoSmithKline, Novartis and AstraZeneca. Inside, a team of young scientists engaged in various tasks, wearing white coats and face masks. These syntheses are not complex, given the correct equipment, a graduate-level organic chemistry education, and easy access to precursors and reagents. The laboratory was filthy and disorganized, with the standard lab equipment found in any legitimate chemical firm, including vacuum pumps, heat-plates, reaction cylinders and drying facilities.

Parry reported:

Young, rich and brimming with energy, Eric embodies the entrepreneurial spirit of modern China. He sits at his desk beneath a cabinet of spirits and cigars that he dispenses liberally to his overseas clients while secretaries totter in and out carrying samples and price lists.

Eric wears designer clothes, drives a Buick SUV and works such long hours his wife moans that he treats the luxury villa where they live like a hotel. But, for all his infectious charm as he chats and jokes in pithy English in his office in an upmarket Shanghai apartment block, there is a sinister side to the business that has made this chemistry graduate conspicuously wealthy aged 35.[5]

Sipping on a Red Bull – his only vice – the chemist said to Parry, 'I have no time for holidays . . . I have a lot of business on my hands. I need all the energy I can get.'

Uncle Fester, aka Steve Preisler, one of the USA's methamphetamine pioneers and the original narcotic folk devil of the pre-internet age, has kept up on developments in the trade, and says Chinese outsourcing was a logical step for US-based drug manufacturers. 'The cooking of the materials has been outsourced to China because it is impossible to do it here,' he told me by email. He continued:

> These materials are too complicated to be cooked up by a basement chemist and they require access to precursors unavailable to US-based cooks . . . Even if it might be quasi-legal, the cops would just simply be crawling all over them, and if nothing else putting them in jail for violations of hazardous waste laws or [health and safety] violations in their shop. They don't have to worry about that in China, and it's nice work if you can get it . . . The folks on the US [and EU] end are more properly described as marketers. They have little to do with the production end other than picking out exactly what material they would like produced.

Preisler did have concerns about the risks to users. 'I do agree that these designer analogues are quite dangerous, more so than the already illegal drugs they are designed to mimic or replace,' he wrote. 'The banned drugs have quite a history of human usage, and so their side effects are known and long-term usage of them has already been experienced. This is not so with the designer variants. It's all experimental and the users are doing the pioneering.'

Soon after our report, Eric started to sell exceptionally dangerous and novel chemicals under the guise of other, newly popular drugs and word spread around the web that he was not to be trusted.

Chinese chemists profit from the differences in international drugs legislation – at the time of our report for the *Mail on Sunday*, none of the drugs we ordered was illegal in China, and many hundreds of designer drugs remain legal there. Even today, China's drug laws lag far behind those of Europe and the US, with enforcement there focusing instead on the drugs most popular in China itself – methamphetamine, heroin, ketamine and, increasingly as China modernizes and its economy and middle class grow, Ecstasy. And whereas in the past the custom synthesis market was small, underground and operating in the shadows of the web, now, with the growth of the trade and of ecommerce, many more are willing to carry out the work without qualms and openly.

While Chinese chemists innovated, accessing vast databases of scientific literature and anecdotal evidence reported online, as well as books like *TIHKAL* and *PIHKAL*, governments Europe-wide chipped doggedly away at the newly exposed narcotic rockface with their blunt, broken and rusty tool of prohibition. Ireland closed hundreds of headshops, and Reuters reported in May 2011 that shop owners in the recession-hit country would face life imprisonment if they sold substances with 'psychoactive effects'. (Originally, headshops there and elsewhere had sold pipes, bongs, and other cannabis-related paraphernalia. The emergence of mephedrone and other legal highs soon saw chemicals jostling with rolling papers for shelf-space – a picture that can now be seen worldwide. The nascent trade in legal highs had already disturbed Northern Irish paramilitary groups seeking to protect their monopoly on the regular drugs trade, with punishment shootings meted out to headshop owners.)

Scandinavian countries clamped down on mephedrone, dozens of related cathinones, and many of the fake marijuana compounds in 2010, and new laws were also passed in most European countries banning many of the new drugs that had appeared in the previous two years, though there was no central ruling as there had been on mephedrone. Poland shut 1,200 headshops and closed down hundreds of websites with crude laws prohibiting the sale of 'substitute drugs'. Days before the law was enforced, queues were fifty-strong in many stores.

In August 2010, the UK government decided to tackle the market in new drugs by drafting new laws. James Brokenshire MP, working then as minister for crime prevention in the new coalition government, and now minister for crime and security at the Home Office, announced that the Misuse of Drugs Act 1971 was to be amended once more. The new law would not be enacted for over a year. On 15 November 2011 the Police Reform and Social Responsibility Act would introduce the government's flagship response to the new drugs problem: new measures known as Temporary Class Drug Orders (TCDOs). These would allow the British government to temporarily schedule any new compounds that came to the attention of the ACMD – or the media – for a period of a year while the ACMD analysed the dangers and pharmacology of the new substances. 'TCDOs enable the government to act faster, on consideration of initial advice from the Advisory Council on the Misuse of Drugs (ACMD), to protect the public against emerging harmful new psychoactive substances while full expert advice is being prepared,' the Home Office announced.[6]

The new law would not specifically name any novel substances; its purpose was political. It was designed to enable the government to react decisively and quickly to the appearance of new drugs without further weakening the authority of the ACMD, which had been so summarily mistreated and ignored over the mephedrone law change.

But between the mephedrone ban in April 2010, and the enforcement of the new TCDO law in November 2011, many new drugs appeared. Alpha-methyl tryptamine (AMT), a psychedelic stimulant that fosters empathy and euphoria, had originally been prescribed in Russia in the 1980s as an antidepressant under the brand name Indopan. In 2010, post-mephedrone, it was sold in the UK, again as plant food. With deadpan brilliance, it had been noted by Russian researchers that the drug, widely prescribed to the elderly, caused inappropriate smiling. However, the drug never gained the kind of popularity that mephedrone did, since its effects lasted twenty hours, while most recreational users are looking for a shorter high. What's more, some users suffered stomach cramps and diarrohea. A final death knell may have been its smell: occasionally an impurity produced in a synthesis of the drug, using indole3-carboxaldehyde as a starting material, forms a compound known as skatole, or 3-methylindole, which is the chief fragrance in mammalian faeces.

AMT is still, at the time of writing, legal in the UK, since the compound has not been specifically targeted by name or by definition. Whereas many ring substitutions of tryptamines had been outlawed in the UK in the Misuse of Drugs Act in the 1977 amendment that so confused the House of Lords, AMT, with its simple structure, was legal. It still is.[7]

Notwithstanding the problems mentioned above, dozens of sites selling AMT sprang up in summer 2010, many of them owned by web-and-drug-savvy individuals who knew AMT had slipped through British drug laws. Many people enjoyed the high, saying it was similar to MDMA and LSD, but with a longer duration than both, and with longer-lasting energy for users. The typical dose range is around 40–50 mg, and users must avoid all other drugs and alcohol if they take it. (Some drugs and even some foods contain chemicals known as monoamine oxidase inhibitors, MAOIs. When taken in conjunction with drugs such as MDMA, AMT, or many hundreds of other drugs that act on the serotonin system, MAOIs – which can be found in hundreds of legitimate medicines and other research chemicals – can amplify their effects to a dangerous level, causing serotonin syndrome, psychosis, or hypertensive crisis and possible death.)

Some sites sell AMT in so-called pellets, to avoid the punishments they would face under the UK's Medicines Act if they sold it as pills – but even at the time of writing they have little to worry about, since not one site has been targeted by British police, who are as confused as users as to what is illegal, real, banned or bunk.

Many other categories of drugs have appeared for sale online since mephedrone was banned. Tranquillizers, long available only to those holding a valid prescription due to their abuse potential and addictive nature, became widely available when a legal, diazepam-like drug, phenazepam appeared for sale on-line on sites that had previously sold only mephedrone. It was legal as it was not included in the 1971 Misuse of Drugs Act – unsurprisingly, as that law had been written four decades earlier, and the drug was a little-known Russian epilepsy treatment. Before the web and social media spread word of its use, it was neither available, used, nor known about, and therefore never deemed to be a danger to British people. Its legality shows that the UK's drug laws are now hopelessly anachronistic.

It only cost about two pence for a dose, and many would be grateful for the effects of the drug, especially after a night spent abusing stimulants. However, many dosed the drug inaccurately – hardly difficult, since the active range for the chemical is just 1 mg – and its side effects were wildly unpredictable, with some users reporting days of narcotic blackouts. It also has a half-life of sixty hours, meaning it remains present in the body for a much longer time than most safer illegal or scheduled prescription-only drugs. And because of the amnesia the drug induces, people often simply redose as they forget they have already taken it. The impacts of all of this are hard to measure, but in July 2011, Dundee University researchers told the BBC that the drug had appeared in nine post-mortem cases, though evidence that the drug killed the users was inconclusive.[8]

Phenazepam was banned on 13 June 2012 in the UK[9]; within a few short weeks, another tranquillizer, etizolam, was on offer online and in headshops across the UK and Europe, sold as plant-feeding 'pellets', again to avoid medicine and food laws. The powdered version of etizolam also appeared – and its dosage level at just 1 mg meant that 1 gram costing about £40 contained 1,000 doses – or four pence per dose. Etizolam, like all those in its class, including Valium (diazepam), phenazepam and Xanax (alprazolam), is addictive and its withdrawal symptoms are more harrowing than those of heroin, and actually more dangerous, since a non-tapered, sudden cessation of these drugs in an addict's body can lead to seizures and fits. It was, as ever, sold marked 'not for human consumption'.

Etizolam was not illegal under the UK Misuse of Drugs Act, nor was it controlled under the Medicines Act, since at the time the law was written in 1971 it was not sold commercially, but most crucially of all, because at that point the net did not exist and nor did this culture of legal highs or research chemicals sold over it. Etizolam is still legal at the time of writing, and is regularly sold on eBay.co.uk.

4-AcO-DMT, illegal in the UK, but unscheduled in the USA, Spain and some Eastern European countries, started to become more widely available that year, too, and won many devotees who found its action similar in every way to magic mushrooms – and indeed it was, for it had been designed as a substitute for psilocybin in official medical tests, again in the lab of Professor David Nichols. Potent and active from around

5 mg, with a standard dose of around 10–18 mg, it required users to buy specialist milligram scales before taking it. There are very few negative reports about the drug online, and some glowingly mystical experiences have been reported. The origin of the drug was a 1999 paper by Nichols entitled 'Improvements to the Synthesis of Psilocybin and a Facile Method for Preparing the O-Acetyl Prodrug of Psilocin', in which he attempted – and succeeded – to find easier, higher-yielding methods of producing psychedelic agents for use in laboratory research into receptor binding properties, structure-activity relationships and the search for new anti-depressant medicines.[10]

Yet more new drugs appeared. Compounds such as 4-HO-MET, another mushroom-type drug, came on general sale. Users reported that it gave them bright, colourful, cartoon-like visions, and plenty of energy for dancing. Then there were new amphetamines that had a flourine molecule bolted onto the three- or four- position of the carbon ring and which gave users the hit of quality cocaine. They were illegal in the UK since our laws forbid ring substitutions of amphetamines, but they existed in a grey area in the US and other countries. Legality aside, they were online and deliverable anywhere in a business envelope in a matter of days.

More new synthetic cannabinoids appeared, both legal and potent. But the real search, in the UK at least, was for a new mephedrone, a euphoric stimulant active in medium-sized doses. When word of a new, legal analogue of MDA, known as 6-APB, hit the web in 2010, things gathered pace once more. MDA itself is an analogue of MDMA, and is illegal, since its structure had been included specifically in the Misuse of Drugs Act in the 1977 amendment. MDA had appeared briefly in the 1960s in the US when it was known as 'the hug drug' or 'the Mellow Drug of America'. Its effects were much like MDMA, though a little speedier and much trippier.

According to data from the Swiss database Ecstasydata.org and dozens of Dutch testing centres, MDA had hardly been seen for almost two decades on the British and European black markets, since the Snowball phenomenon of the early 1990s, though supplies in the US seemed to remain fairly constant. The reason for the prevailing shortage of MDA are twofold: first, the drug is not as easy to use as MDMA, since it is longer-lasting and more truly psychedelic, with startling hallucinations, as detailed in earlier chapters, frequently reported. Second, chemists making MDMA will get a greater yield of that drug from their expensive and illegal precursors than they would if they aimed for MDA. Therefore, in the web drug subculture – in the new legal highs mainstream – a legal MDA analogue was major news.

Previously, there was just one report online of an MDA analogue, called 6-APDB, by a poster named fastandbulbous at the Bluelight forum, back in 2006, when the research chemical scene existed as a hidden subculture following the high-profile busts in Operation Web Tryp. Fastandbulbous had access to the very rarest of research chemicals and his reports were clear, Shulgin-esque affairs often tinged with an amusingly eccentric hippyish sensibility – as you might expect for a man whose online name takes its inspiration from a Captain Beefheart track.

6-APDB was illegal, because its chemical formula had been covered in amendments to the Misuse of Drugs Act. Not that fastandbulbous paid heed; his appetite for psychedelic and narcotic novelty was limitless and he didn't much care about legality. 'The effects [of 6-APDB] are similar to those of MDMA . . .' he wrote. 'There is increased empathy and profound contentment, and a luxurious sense of tactile enhancement. However, unlike with MDMA, there is no drive towards speech or locomotor activity – even though I would not characterize the compound as sedative in any way.'

Another poster, robatussin, replied portentously: 'I'm glad there are people out there that know what they're doing, living with Shulgin's spirit! The government will have to outlaw basic elements like nitrogen or oxygen soon to keep pace with chemists.'

By 2010, four years later, there were many more people reading Bluelight and looking for drugs online than in previous years, because mephedrone had changed the drug culture so dramatically and so publicly. That summer, someone somewhere read fastandbulbous' curious narcotic memoirs and arranged for a legal version of 6–APDB – called 6–APB – to be synthesized in a Hungarian laboratory. A tiny modification to the MDA skeleton was made by removing one oxygen molecule and double-bonding part of the secondary carbon ring, as can be seen in these diagrams:

*MDA*

*6-APB*

This small tweak had changed the drug's chemical categorization; it now had a benzofuran ring, and therefore belonged to an entirely distinct chemical class, making 6-APB slip through the law's loopholes in most countries in the world, including the UK. A new, completely legal drug had appeared. A poster named GZero wrote a trip report on this new drug on 7 May 2010: 'This is an absolutely beautiful drug . . . It's like MDMA but less sweating fiendishness. Cocktail party MDMA. Very similar to MDA actually, but gently and colourfully visual. This is insane, I never expected this to be this nice.'

In the past, drugs like MDMA took years, if not decades, to become popular. With 6-APB, it happened in a matter of weeks. A group of five UK-based vendors claimed to have the exclusive distribution rights to the material, which was sold initially at eighty pounds per gram – making each dose cost around twelve pounds. They posted YouTube videos, now deleted, of the pill-pressing and packing processes, attempting to brand the drugs exclusively. They also posted videos showing the correct reaction for the drug using widely available pill-testing kits.

Most brazenly of all, the drug was branded Benzo Fury by the vendors. It was an odd choice of name because 'Benzo' is usually a slang term for tranquillizers, taken from the first syllable of 'benzodiazepines', the class of drugs to which diazepam (Valium) belongs, whereas this new drug was nothing of the sort; it was a psychedelic stimulant. But there was a chemical connection: a benzofuran is a benzene ring – a circular structure made up of six carbon atoms and six hydrogen atoms – that is connected to a furan ring, which is made up of four carbon molecules and one oxygen molecule. And in the same way that 3,4 methylenedioxymethamphetamine only really took off when it became known as Ecstasy, and just as 4-methylmethcathinone became much more popular once it was called Meow, so too did 6-(2-aminopropyl)benzofuran, aka 6-APB, gain many more fans when it became branded as Benzo Fury. It was sold from that summer of 2010 right up until this book was going to press, in professionally printed orange plastic pouches, complete with a barcode. (Official though it looked, however, the barcode on many packets actually corresponded to the 1986 Genesis album *Invisible Touch*.)

With their branding exercise, the vendors had also chanced upon a singularly effective web marketing strategy. When people entered the two search terms BENZO and FURY into Google, the sites owned by those selling this new drug appeared first in the search results, since it was such a strange and unusual phrase. They also tagged their sites' and pages' metadata – information which Google's robots crawl, document and then index – with the chemical and brand name of the drug. It was the finest bit of designer-drug-related rogue search engine optimization that had ever been seen. That was, admittedly, a very small field – at the time. It has grown since.

Almost as soon as the trip reports by GZero were published, dozens of European and British outfits claimed to have produced the compound. Some Chinese manufacturers immediately claimed they too had produced it, circulating fake documentation, and then sending dozens of kilos of fake chemicals to vendors all over Europe. Vendors sent these compounds out without testing them; dozens of people fell sick and at least one man suffered a seizure. But within months, a large number of vendors had started selling pure 6-APB to happy customers. It was popular and legal in the US and most of Europe – or more exactly, it was not specifically scheduled, and remains so even today. Nor has it been targeted under the UK's TCDO law.

With breathtaking speed over the next two years, hundreds of new compounds came onto the market, mainly sold by UK-based websites that had started out in the mephedrone trade. The research chemical and legal highs market, already a confusing cornucopia, became even more atomized as vendors clamoured for products that would satisfy the search for novelty, legality and profit. Now, instead of selling one or two compounds, sites were selling, or claiming to sell, dozens. All of the banned drugs that were swept out in the cathinones and JWH-series ban of 2010 were swiftly replaced. In place of JWH-018, the synthetic marijuana replacement, there came AM-2201 and AM-2203 – both untested, and stronger than the drug they replaced. It was like a game of chemical musical chairs. Camfetamine, a legal analogue of banned appetite suppressant fencamfamine came that year, followed swiftly by methiopropamine, a crystal methamphetamine analogue. Ethylphenidate, a legal analogue of ADHD medicine Ritalin, was next. And the vaults of Shulgin were plumbed once more as vendors found one or two oddities from *PIHKAL* and *TIHKAL* that had not yet been banned. Dissociatives, drugs in the family of ketamine and PCP, or angel dust, came online soon after, and the next break-out legal drug – methoxetamine – was born, with the first trip report posted online in 2010.

Ketamine (also known as K) is used in medicine as a general anaesthetic. It is favoured for its ability to keep patients' airways open, unlike other anaesthetics, so tends to be used in accident scenes such as car crashes and battlefield amputations where there is no oxygen available. Ketamine is also the only anaesthetic available that is pain-relieving and sleep-producing. Inquisitive drug users took it, perhaps inspired by American neuroscientist and polymath John Lilly's experiments with the compound in the 1960s. He often took it in sensory-deprivation flotation tanks in a bid to explore the outer limits of human consciousness – and to treat his lifelong affliction with migraine headaches. (Lilly was also fascinated by dolphin-human communication and the study of extra-terrestrials, and he later psychotically claimed that he could communicate with alien intelligences representing good and evil, and constructed a personal cosmology he called ECCO, Earth Coincidence Control Office.)

Ketamine is tailor-made for those on a mission of complete

personal oblivion. Its import in the mid-1990s when it first appeared on the drug scene, was often carried out on a small scale by hippy travellers smuggling it from India – where it was legal and widely available at the time – inside one-litre bottles of rosewater laced with an average of 50 g of the drug. It sold for between £200 and £600 per litre, extending by many months the gap-year rupee-whoopee of those who sent it home. Between the mid 1990s and 2010 ketamine was available in great abundance in the UK and mainland Europe, and very cheaply, at ten to twenty pounds per gram. At first it was popular at squat parties and warehouse gatherings, and it was associated with the seamier end of the rave scene. Some clubbers derided its use, a stance easy to sympathize with once you have seen a user staggering clumsily in an anaesthetized haze around a dancefloor.

The 2010–11 British Crime Survey, published by the Home Office, found that 714,000 people are estimated to have taken ketamine in their lifetime, with 207,000 using the drug in the last year.[11] In 2006, the drug was made Class C in the UK, but the government's scientists did not look ahead and ban all of the possible variants of the drug – that is, they did not impose controls around ring substitutions on the basic skeleton. This was not neglect on their part; rather, there was simply no precedent at that time. And because there was plenty of ketamine, nobody bothered to innovate. But at the start of 2011, Indian authorities put in place new rules banning the drug, and what had been a dry patch became the UK's first full-scale, year-long ketamine drought with prices tripling from ten to thirty pounds a gram.

News of a legal ketamine analogue had first surfaced in May 2010. The drug was invented by Karl, a clandestine theoretical chemist, to whom the tangled drug laws pose an intellectual challenge. I met him on 18 April 2012 in a grim pub in a town in the north of England. Karl is an obsessive genius who knows more about drugs than most paid experts, though he doesn't work. He has a degree in biochemistry and a parallel, lifelong fascination with altered states, and has unearthed, designed or conceived of – and consumed – more drugs than most politicians even know exist. He is personally responsible for the emergence of two of the most popular drugs that followed mephedrone, 6-APB and methoxetamine. He knows of dozens more, all of them sitting outside most drug laws in most countries. For now, he's keeping most of their structures secret.

Diffident and mild-mannered, he's not a dealer, and he's not a chemist, and he doesn't make money from the trade, nor does he want to. His work is theoretical, an intellectual endeavour located at the edges of science and hedonism, and the law doesn't interest him, except insofar as it guides his investigations. Karl says the UK ban on mephedrone and other ring-substituted cathinones actually missed plenty of alternatives. 'They think they banned all the viable structures,' he says, 'but they haven't.' Each of the sentences he said next listed structures that, if commercialized, would be worth several hundred thousand pounds – if not more – and many of them already have been.

Karl has suffered from phantom limb pain since blowing his hand off in a teenage accident, and he knew from researching medical literature that ketamine could help. But he found certain aspects of the ketamine experience traumatic. Such as? 'Alien abduction,' he says, deadpan. There were other problems, too: the drug is addictive, and its metabolite, norketamine, causes damage to the urinary tract so severe that some sufferers have had to have their bladders removed. Around 2009, Karl set about a trawl of the scientific literature and designed methoxetamine. 'I read research papers and found that changing the substitutions of the molecule would take away the more bizarre elements of a K experience,' he said. 'I wanted something less deranged. So I invented methoxetamine. It's basically ketamine, but instead of having a 2-chloro group on the phenyl ring it has a 3-methoxy,' he said. The new drug Karl invented can be seen opposite, along with the ketamine molecule it was based on.

*Ketamine*

*Methoxetamine*

Methoxetamine was legal, and dislocated the mind and body just as well as its parent compound. It was initially sold online at around seventy pounds a gram, but it was active at far lower doses than ketamine. One gram of ketamine could get two or three people high. One gram of methoxetamine could get up to 100 people high. The price dropped within a few months, making it even better value for money. Response online was immediate and positive.

Soon, there were dozens of sites selling the compound – some of them blatant enough to call it 'K's Sister' – and it was possible to email chemical factories in China and have a kilogram of the drug made for around £7,000. Some sites selling methoxetamine, in common with the previous mephedrone traders, paid for Google's AdWords programme and so searches for 'legal highs', 'methoxetamine' or the name of the chemical compound yielded paid-for links to their sites, and Google once more profited financially from the trade in semi-legal drugs.

Methoxetamine was selling quickly and winning thousands of users, some of whom had been drawn only by its legality, some of whom were interested in trying any new drug experience simply for the buzz, while others yet were looking for a ketamine replacement. The *MixMag* survey in 2011 found that almost five per cent of its 7,000 respondents had tried methoxetamine, and the magazine ran a feature in its January edition outlining the dangers – and pleasures – of the drug.[12] Three hundred and fifty users may not sound like a great number, but consider the number of people first arrested in Operation Ismene – the British counterpart to Operation Web Tryp: just twenty-two. And *Mixmag*'s question applied to just one drug, and a strange and unusual one at that – dissociatives are already a fairly niche interest within the drug culture. Within a decade, the market in incredibly obscure research chemicals had grown, by this measure, by a magnitude of fifteen.

But those figures only provide part of the picture. In conversations with laboratories serving the UK, I have established that several hundred kilograms of methoxetamine were imported into the UK between 2010 and 2012. Each gram can deliver 100 doses, each kilo 100,000 doses, though it's not always used at such low levels – some people take very large amounts as tolerance builds. Research into the drug has been limited and evidence is at best anecdotal, but between 2010 and 2012 it was undoubtedly extremely popular with young people short of cash and contacts, who found the potency and legality of the drug a major draw. Tales from users, like those from all users of dissociatives, were suitably bizarre – those in the *MixMag* feature reported imagining themselves as snowmen in Kate Bush videos, or thinking they were on board canal narrowboats when they were in fact in the Shacklewell Arms, a grubby hipster pub in east London.

At 10 a.m. on 28 March 2012, the government announced that methoxetamine was about to be placed under temporary control order in the UK. The ACMD had become aware of the drug, and had alerted the government to its emergence and use. The drug had not captured the imaginations of newspaper editors quite as mephedrone had, mainly since it was never as popular. But there were a number of deaths from the drug, details of which follow in the next chapter. Sites started selling off stock at £4,500 per kilogram – a £64,500 per kilo discount from the profits made in the early days, but still a decent profit, as wholesale prices had dropped to around £3,000 per kilo. By 11 a.m. that day, sites had removed the chemical from their menus.

But legislators again failed to look at the drug's chemical structure and fully understand the family that this new compound belonged to. When methoxetamine was temporarily banned none of the more than fifty possible psychoactive variants to the arylcyclohexylamine skeleton were included. The day of the ban, sites selling tiletamine and 4-MeO-PCP, two other ketamine analogues belonging to the PCP family, started to jostle their way up the Google rankings by including the meta-tags 'methoxetamine' in their page descriptions and optimizing their pages for search. By late 2012, another dissociative ketamine substitute, N-Ethyl Ketamine, was on sale at dozens of sites.

When the ACMD returned its final findings on methoxetamine in October 2012, it advised the government to outlaw the drug outright, since it bore such a close resemblance to ketamine in its harm profile. It recommended that many other – but not all – named derivatives of ketamine be outlawed, too. Ketamine and its derivatives, including methoxetamine, are all addictive, or at least habit- and tolerance-forming, meaning larger quantities are needed to get the same effect if the drug is used regularly. They also cause short-term memory loss, and, as mentioned above, ketamine's metabolite, norketamine, can cause severe damage to the urinary tract. One of Karl's aims when inventing methoxetamine was to reduce the physical dangers of the drug; the reports from the ACMD suggest he failed.

Around the time mephedrone first appeared in 2008, I started to investigate the chemical underworld in more detail. One site in particular seemed emblematic of the way new drugs were being bought and sold. Right up until September 2010, this site was the subject of constant online chatter, with customers raving about the compounds available, the variety, the innovation and the market-beating customer service. It closed suddenly and without warning in September 2010, and its absence left a large and lucrative gap in the market that many other operators rushed to fill over the following years. It had been the biggest retail supplier of research chemicals the web had ever seen. The size of this one business – which had a turnover of around ten million dollars a year for the previous two years – is a good indicator of the size of the market, which has only grown since.

In 2011, I spotted a new site whose name bore a certain resemblance to that of the closed store, and guessed, correctly, that the old site's owner had moved out of the retail business into the wholesale market. His name, for the purposes of this book, is Matthew. We exchanged several emails, long, thoughtful affairs rich with detail and insights into the business.

Matthew says his business was, at its peak, shipping around a hundred orders worldwide daily, from his fulfilment centre in Taiwan. The compounds were sent in standard envelopes often decorated with Hello Kitty stickers and logos to make them look like birthday cards or love letters. Inside were research chemicals such as 2C-P, 2C-E, 2C-I (Shulgin's psychedelic phenethylamines from *PIHKAL*), and 4-AcO-DMT (a drug identical in many ways to magic mushrooms, from *TIHKAL*), and dozens of others. His store was so well stocked and well managed that it became known as 'Wal-Mart'.

'I had staff [in Taiwan] working nine to five weighing and bagging the product, and others stuffing envelopes. All manual labour. We had in place a pretty rigid set of QC checks to prevent any errors. In the back-end, I built a custom secured database for maintaining inventory and customer records,' he said.

Matthew was an early entrant in the massive market for mephedrone. From the outset, he was concerned with quality. 'Early on, all available mephedrone was very poorly synthesized, and had a terrible shellfish odour due to the fact that the demand was so high and the Chinese labs were producing it so quickly that they were not completing every step, or not letting it dry completely before packing and shipping out. I had many heated words with my labs about that, and I went to several labs before finding one that could produce a quality and consistent product. By that time, however, it had been banned in the UK and the trade volume was winding down.'

He says he speaks reasonable Mandarin Chinese, having lived and worked there for over ten years previously, and he used labs in China, although many of them did not have the required pharmaceutical expertise to produce the drugs he and his customers wanted. 'If you're usually making melamine or some other industrial chemical you don't need to know, let alone meet, pharmaceutical standards. Of course people were not actually feeding their plants with this stuff, so there was a lot of concern on my part about these labs. I visited one lab which stank so bad and was so dirty I wanted to puke when I stepped inside. I immediately stopped working with that lab after seeing the actual premises.'

Business was brisk. The firm was shipping 700 orders a week, usually containing three different products on average – that is, it was fulfilling over 100,000 individual drug orders annually, and it operated for more than three years. Many of the drugs Matthew sold, if not the majority, are active in the single-digit milligram range – meaning his shop alone shipped tens of millions of doses of drugs each year. The research chemicals market was now mainstream and globalized, and the profits were even more mindblowing than the drugs: 'We shipped pretty much everywhere. The usual mark-up at retail was 3,000–4,000 per cent. The company netted several million dollars per year, right off the bat. It took off real quick. The website was closed to new customers most of the time because we couldn't handle any more volume. The bulk of the business was from our return customers. If we'd been able to open to new customers we could easily have doubled those figures.'

A web professional, Matthew says his company was successful because he had previous professional experience in mail-order customer service, internet marketing and web programming. Before his site, vendors had used email only or had only rudimentary 'free' websites. 'I introduced the "shopping cart" to retail research chemical sales,' he says. 'Starting with some open-source shopping cart software, I personally wrote the payment API modules and many other custom functions for the website. Now you could complete your order, and pay for it (using AlertPay) within minutes, and you would get automated emails updating you with the status,' he says. Matthew's site set the technical standard for those that would follow.

The web enabled Matthew's business to operate remotely, and he took a hands-off role much of the time. 'I built an offshore fulfilment center as soon as I could not handle the load by myself (which was quite quickly). We had the fulfilment center in Taiwan, labs in two or three other countries, I was in yet another country, and often traveling the globe for various reasons, yet everyone was interconnected and you could often complete your order and see it ship within two hours,' he explained.

He also sold to other retailers worldwide as he built up a large and varied stock very rapidly. At one point, he was selling over twenty different products, most of which would be unknown to most governments and customs forces worldwide – but which were the subject of hundreds of threads on drugs forums. Just ten years previously, most drug users could choose from heroin, cocaine, LSD, marijuana and hashish, and amphetamines or Ecstasy. Matthew sold five different kinds of synthetic marijuana, a heroin analogue, several hallucinogens, three or four versions of drugs that were like cocaine and many more.

He was an early adopter of Twitter, using the microblogging site to inform customers of sales and offers and promotions, and was one of the first people in any trade to use QR codes – the scannable, almost bitmapped black-and-white icons that, when scanned and decrypted, would send customers to secret URLs on his site that had special deals. He remained sober while working, and ensured that all of his staff did too. His own drug habits while not working are a topic he chooses not to discuss.

The real enabler in the whole trade, he says, was the willingness of Canadian e-wallet firm AlertPay to process credit card payments for these drugs. 'If you want to know what enabled the RC revolution from 2008 to present, it was AlertPay,' says Matthew. 'They've basically chosen to look the other way regarding these transactions. AlertPay had a banned product list, and still maintains one, but it was a random mess and did not relate to any laws, and as there is no global treaty on these compounds, it wouldn't have made any difference in any case. Without AlertPay, the RC business absolutely could never have taken off like it did. [AlertPay] made millions in profit directly from the RC business.'

After a question was raised about the firm in the House of Commons in 2009, AlertPay shut its dealings with some sites. The family of Max Llewellyn, an eighteen-year-old Welshman who hanged himself after becoming depressed following a mephedrone binge, spoke to AlertPay in March 2010, asking them to stop processing payments for mephedrone vendors. Its business operations director at the time, Elsworth Weekes, insisted the company did not endorse the misuse of mephedrone, which was legal at the time. 'Should the government of the United Kingdom render a decision that should affect the legal status of mephedrone, AlertPay will take appropriate initiatives to make sure that its services remain in total adherence to trade laws,' she told a local Welsh newspaper.[13]

AlertPay shapeshifted in May 2012 to become Payza, and now says it will not take payments for 'Drugs and related paraphernalia (including but not limited to research chemicals and illicit herbal incense)'. Today, the Natwest processes bank transfers for the biggest online research chemical company in the UK, whereas PayPal, the net's most favoured online payment processor, does not allow the sale of research chemicals via its service, and seizes funds without release if vendors are found to be breaking the rules. Some sites today get around the problem by simply using multiple accounts. But the fact that PayPal can adequately manage to prevent the sale of these chemicals through its services shows that it is possible, with will and vigilance.

Matthew says he manned the site remotely round the clock. 'I ran the business like a professional, not like a shady criminal. I was also principled and honest, perhaps the very opposite of what people had dealt with up to that point. Anyone could have done what I did, I just happened to be the first and I guess I benefited from that to some extent.' Money was never a motivation; rather he claims he was a chemical evangelist: 'I just wanted to share this stuff with the world, but then it quickly got out of hand. This was really just a hobby that I eventually got bored with. So maybe that helps you understand a little bit why I abruptly shut it all down, right at the peak of our business, without any apparent concern. As it turned out, I think I quit at the right time, as the RC industry took a turn for the worse soon after we were gone.'

★

In the years following the ban on mephedrone, for every forum that started, another closed or rebranded following in-fights or buy-outs, in a process that reflected the ever-warping world of new drugs. One site, Legalhighguides, was populated mainly by American teenagers, judging by the content and tone and linguistic patterns. It was very busy, with thousands of users posting hundreds of threads every day. The site had a popular 'swap-meet' section, where users could offer each other trades or sales of drugs. It was quickly filled with vendors and Chinese laboratories and others selling new designer drugs. The conversation was global and public, and soon dealers sponsored it, handing out free samples of drugs in exchange for adverts. Everything from Shulgin's chemicals to synthetic cannabinoids and a wide array of other chemicals were available, and no one cared about legality.

The site shut down not long after one dealer duped dozens of users out of many thousands of dollars, and shortly a competing site launched, The Euphoric Knowledge. Its server was hosted in Holland, and soon enough dozens more group buys and swaps and sales kicked off. The site was mainly populated by young men who bought and sold designer drugs to each other with an enthusiasm and blatancy that was matched only by their carelessness. This Facebook generation, so accustomed to sharing information openly and indiscriminately in a frictionless world where an acquaintance or an adserver alike are trusted with access to your most private information, was lulled into a false sense of security by the lack of action in the US in the years following Operation Web Tryp – in the unlikely event that they had ever heard of it.

Quite how anyone believed that a site such as The Euphoric Knowledge could continue to run is anyone's guess, but for a while, it was one of the busiest spots for the research, purchase and sale of some extraordinarily rare and potent compounds that, just a few years before, were known perhaps to a few thousand people worldwide. The users of the site offered hundreds of different chemicals for sale, with an almost unquenchable appetite for novelty. They believed they would not be prosecuted under the American Analog Act for chemicals that had not yet been listed specifically as illegal, pinning their hopes on the 'not for human consumption' defence. They created thousands of threads in which they discussed openly the price for bulk import and export of chemicals that most American judges would, in a heartbeat, class as illegal analogues of banned substances. It was only a matter of time before the axe fell.

The site's founder, Justin Steven Scroggins, known as w00t, was arrested on 10 April 2012. Undercover federal agents had infiltrated his site – not a hard task, since registration was open – and had eavesdropped on his Skype calls with a laboratory in the Jiangsu province of China that is still operating today. At the time of the investigation, this laboratory sold only seventeen products, all of them considered analogues of banned Schedule 1 substances in the US, according to the indictment. Scroggins was charged with the importation of just over two kilos of cathinones, none of which was specifically illegal at the time, but all were considered analogues of methcathinone and other banned substances. He pleaded guilty and was awaiting trial in late 2012.

The research chemical market is now in the hands of thousands of sites that sell hundreds of chemicals to unknown numbers of users. The sites' owners run the gamut from paranoid American teenagers, to idealistic Spaniards, or Chinese labs that have now opened retail sites and are happy to send potent white powders out to anyone with a Western Union payment slip. The blurring of the lines between legal highs, research chemicals and illegal drugs complicates an already labyrinthine legal and chemical picture, and poses a dilemma for legislators, users and law enforcement agencies.

How easy is it to circumvent British and other drug laws as

they currently stand? 'It's simple, laughably so,' Karl, the inventor of methoxetamine said when I met with him. Can he think of a cocaine analogue that might be outside the law? Instantly he replies with a chemical name and formula. 'It's active at the same potency as regular cocaine. It has the same receptor binding affinities and reuptake inhibition as cocaine.' Has he tried it? A nod. 'The only trouble with this compound is that it requires ecgonine as a precursor, and that's a controlled substance. But someone could find a synthetic route to making that, there's no reason why not,' he says.

How about a legal heroin analogue? 'There are dozens.'

Karl has also read the works of Alexander Shulgin, and has conceived of ways to take those now-banned compounds and make them legal. He then reels off an unintelligible string of multisyllabic chemical terms that only chemistry graduates would understand. 'There are some in there where you could replace the methylenedioxy ring with a straight three-carbon bridge – rather than an oxygen-carbon-oxygen array, you could have three carbons. By replacing the methylenedioxy group on some of the *PIHKAL* entries with a trimethylene they would be taken outside the 1971 Misuse of Drugs Act, and they would be active,' he says.

Karl's story, though perhaps one of the most extreme in this emerging underground, is instructive. There are dozens of Karls posting online now. Some specialize in heroin and opiates, others in stimulants. Every class of drug has a few experts debating ways to make a legal version of an outlawed compound. All of them are bright and well educated, with the knowledge and contacts to produce new drugs in a matter of months, and anyone can read their comments. It's not so much that they have no respect for the law, it's more that domestically, there's no easy way to write a law that can control their innovation yet still allow legitimate industrial and medical research to continue.

Pull the focus wider to bring the largely unregulated Chinese chemical industries into the shot, and then zoom in on that country's drug laws that limp decades behind the developed world's, and the situation becomes extraordinarily difficult, if not impossible, to manage.

But this is not just a question of legality or enforcement. Safety is the more important issue, and for all the professionalism that people like Matthew claim to have fostered in the industry, accidents in this trade can and do happen. When they do, the results can be serious, if not fatal. In 2010, nine customers of his firm were sent packages containing 2C-P – a Shulgin psychedelic active at 8–12 mg. But the bag was mislabelled as buphedrone, a mephedrone variant with a standard dose of around 80–100 mg. A number of customers took it, and overdosed. One of them twice. Their stories follow in the next chapter.

One site that aims to address the problem of fraudulent or dangerous vendors is SafeOrScam.com. Known as SOS, it is not a place to source research chemicals, or traditional drugs, neither is it a place to sell drugs, or even to talk about them. SOS is an internet drug vendor-review service that tells you whether using a dealer whose details you must already have, is safe, or a scam. The site does not publish any lists, its databases are encrypted, and SOS only gives information on sites you already know the name of. Will vendors deliver the product they advertise, or will they send you a dangerous substitute? Will they rob you of your money? Are they a cover operation for a police sting? The way it does this is neatly collaborative and once more reflects the new primacy of the crowd over the individual.

Users of the site – which is open, on the surface web, and for its most basic and useful function requires no membership – simply enter the URL of a website selling research chemicals and the results come back with two scores. One is a score between one and ten representing the entire period since the site was first entered on the database, and the other is the average of the scores accumulated over the last thirty days, with the number of votes cast also shown. Click through the ratings and you can also check the online reputation of each of those who voted and judge for yourself if they seem truthful. This reputation system for the site's users, known as 'karma', is accumulated over time in reward for useful posting, and points are deducted for inaccurate or dishonest information. While not foolproof, it is a clever way to ensure that the site is not swamped with shill approvals or gamed by unscrupulous dealers. The site, which is ad-free and baldly but brilliantly coded, has saved people's lives.

SOS is in the purest terms nothing more than a network, a platform for information exchange and a virtual gathering place for people who want to share knowledge, for no personal profit. The owner of the site, an American software coder, is an articulate and precisely rational advocate for the free exchange of information, especially around research chemicals and other drugs. The site gets about 10,000 hits a day, and attracts around 150 comments in the threads below each entry. The owner, who calls himself simply Admin, says that comments on the site increased by around sixty per cent in March 2010 compared to the previous month – from 632 to 1,078 – just before the UK banned mephedrone – and plummeted soon after to normal levels. This suggests that many users were stocking up and wanted to ensure they had suppliers that were trustworthy. Visitors to the site and comments there increased five-fold over the following year as the research chemical market grew, he told me.

Like so many other administrators of drug-related sites, Admin is adamant that his site does not encourage or enable drug use. His views could be read as a de facto manifesto for many involved in the research chemical and online drugs scene, who see themselves primarily as information activists rather than drug-law reform campaigners. Rather than waiting for the law to catch up with the growing number of people who believe that drugs should be legal, and that information should be free, the SOS

Admin has simply used information-age tactics to sidestep pre-web laws. He told me:

> I've gone to great lengths to prevent people asking for or giving sources for drugs or research chemicals on the site. I'm not opposed to freely available sources from a philosophical perspective, but from a practical, legal perspective it's not a good idea to make sources widely available. I've been involved with research chemicals for approximately five years. I've always had a fascination with chemicals and the effect they have on the human body. I don't think it's up to the government to judge any person for what they do with their own body in the privacy of their own home. In my opinion, the primary role of the government is to prevent violence. Is the prohibition of a substance absolutely guaranteed to prevent more violence? If the substance is sarin, I think the answer is yes. If the substance is marijuana? Absolutely no. The same goes for cocaine, heroin, 4-AcO-DMT, mephedrone, etc. Information should absolutely, unequivocally be free. It should never, ever be illegal to receive information. It should never, ever be criminal to give information. This is absolutely my most strongly held conviction, and one from which I doubt I will ever stray.
>
> The biggest change with the research chemical scene is that the amount of information available has increased dramatically. Five years ago, the only information you could access about research chemicals was through sketchy forums and chat rooms from individuals who were unreliable at best. Today, Wikipedia (and the linked references) contain a wealth of information on hundreds of research chemicals. If SOS helps people make informed and successful decisions I will be incredibly glad. That is the entire purpose of the site. What they do after the

initial decision is absolutely none of my concern. It doesn't bother me in the least.

The site has reviewed 13,560 vendors, and has 3,086 registered users and over 70,000 comments. Given that it is difficult to obtain a referral to the site that allows users fuller access – only full members can post their comments below reviews – that is an enormous conversation. Consider, too, the concept of on-line participation inequality, or the 'one per cent rule', the social media theorem which states that for every person who posts on a forum, generally about ninety-nine other people are viewing that forum and lurking.

One laboratory in Hong Kong that has a high SOS rating replied to an email I sent them asking what chemicals they sold:

All products are produced in China and ship from China. Shipping with tracking is free of cost to any destination. Shipping to most countries requires about 3–7 days, and orders normally ship within 5 business days of payment receipt. Payment can be by Liberty Reserve (5% discount), MoneyGram or Western Union (please keep receipt).

The products which are currently in stock are listed below. JWH-018; JWH-019; JWH-073; JWH-081; JWH-122; JWH-203; JWH-210; JWH-250; A-796,260; AB-001; AKB48; AM-694; AM-1220; AM-1248; AM-2201; AM-2233; RCS-4; CB-13; URB597; URB602; URB754; UR-144; JTE-907; MDA-19; 5-MeO-DMT; 5-MeO-DIPT; 5-MeO-DALT; 2C-B; 2C-C; 2C-D; 2C-E; 2C-H; 2C-I; 2C-P; 2C-T-2; 2C-T-4; 2C-T-7; 25I-NBOMe (NBOMe-2C-I); Methiopropamine (MPA); 2-Fluoroamphetamine (2-FA); 2-Fluoromethamphetamine (2-FMA); 4-Fluoroamphetamine (4-FA); 4-Fluoromethamphetamine (4-FMA);

Methylone (bk-MDMA); Ethylone (bk-MDEA); Butylone (bk-MDBD); Eutylone (bk-EBDB); Pentylone (bk-MBDP); Ethcathinone; 4-Methylethcathinone (4-MEC); 4-Ethylmethcathinone (4-EMC); Benzedrone (4-MBC); 3,4-Dimethylmethcathinone (3,4-DMMC); Flephedrone (4-FMC); Buphedrone; Mebuphedrone (4-MeMABP); NEB; Pentedrone; alpha-PPP; MPPP; alpha-PBP; MPBP; alpha-PVP; MDPV; MDAI; 5-IAI; Desoxypipradrol (2-DPMP); Dimethocaine (DMC); Methoxetamine (MXE).

Available by custom synthesis is 4-AcO-DMT, 4-HO-MET, 4-HO-DET, 4-HO-MPT, 4-HO-MiPT, 4-HO-DiPT, 4-AcO-DiPT, 5-MeO-MiPT, 5-MeO-AMT, AMT, DPT, DiPT, MET, MiPT, NBOMe-mescaline, 25C-NBOMe, MBDB, 3-FA, 4-methylamphetamine (4-MA), 3-methylamphetamine (3-MA), MMA, IAP, JWH-series, AM-series, URB-series, AB-series and many others.

Regards,

[LAB NAME]

This is the reality of Drugs 2.0. Among that unreadable alphabet soup of drug names there are hallucinogens, stimulants, empathogens and cannabinoids. Working out which of them are legal or which have been outlawed in various countries would require thousands of hours of legal time or case law study.

This lab is trusted; they deliver what they say they will, and they have no interest in what their customers do with their compounds. The old paths between user and dealer, the small networks of individuals with a common purpose and even friendship, have now been replaced for some people by a laptop terminal in a net café in west London, Alabama, Kiev, Scunthorpe or Amsterdam communicating with a nameless encrypted email server in Hong Kong.

And among that plethora of new drugs, as well as fairly benign chemicals, there are others that can kill or injure in seconds if users do not know the right dose. Overlay that fact with this: *MixMag*'s 2012 survey found that twenty-five per cent of respondents were happy to take what it called 'mystery white powders'.[14]

# 8

## *Ready-Salted Zombies and a Chemical Panic*

Addiction, overdose or imprisonment are the guaranteed flipsides to any serious drug habit – and it's proven to be no different with the new drugs created in the last decade.

In the early days of the online drugs scene, there were a limited number of suppliers, a tiny user group, plenty of legal confusion and a good deal of secrecy and prudence. Even then there were deaths. Only after Operation Web Tryp did this subculture become more widely known, and in the eight years since that legal clampdown in the US, the market has grown exponentially. There has been a new influx of users with scant knowledge of the historical scene that spawned the contemporary one. And even as the user group grows, so too does the number of drugs on offer, and the number of deaths.

That said, hysterical media coverage of the perceived threats of new drugs and corresponding knee-jerk government action seem to be similarly guaranteed. Newspapers were, for a few months in 2010, peppered daily with howlingly inaccurate reports of the deaths of Meow-crazed youths. But the truth is that there have only been two specific deaths from mephedrone in the UK, and one of those victims was compulsively injecting the drug intravenously.

John Ramsey believes the picture is more complex than either users or the authorities might admit. 'Most people take more than one drug and some unfortunately die – when they do, it is usually impossible to say which one was responsible for causing the death. Unfortunately the tendency is to attribute the death to the latest drug and consequently overstate the risk,' he told me.

Admin from SOS thinks that the main reason people die from the new designer drugs is a lack of information, and in the US, Europe and the UK that is often true. 'The current market is insanely negative and dangerous. In the US, at least, the way the Analog Act is worded is incredibly harmful. Specifically, according to that law, analogues of scheduled substances cannot be sold for human consumption,' he said. He explains that this means no dosage information can be sold along with the chemicals: 'In the US, the substance is not illegal, and the sale of the substance is not illegal. What's illegal is the *information* that is passed along with the sale – so long as the information includes any details on what the substance is and how to use it correctly and safely. This leads to research chemicals being sold unlabelled and without usage guidelines. This single aspect has been responsible for every research chemical-related death that I am aware of. I think that any adult should be able to go to the pharmacy and purchase any chemical they wish. It should come with detailed usage instructions including side effects, interactions and warnings. The pharmacist should be knowledgeable about these chemicals and should dispense advice to the best of their ability.'

That's a radical and, some might consider, extreme viewpoint. But would we rather have a scenario where people can buy any chemical they want, from Chinese chemists or branded legal high vendors, with no information on dosage and contraindications and no guarantee of purity? For that is the way things are. Would Admin's solution cause more or less harm? It is impossible to say until we try it, or until we devise a strategy that might prevent people looking online for drugs.

Another, more subtle and non-chemical danger is that while

the net may offer a community for drug-users, in itself that can soon become an echo chamber, a hall of mirrors that normalizes once-extreme behaviour. One user told me, 'Of course there are a lot of people deluding themselves that their self-destructive behaviour serves a higher purpose, helps them develop their personality, gain insights or is part of a spiritual path of some kind. Keeping company with a bunch of other out-of-control polytoxicomaniacs bragging about their exploits on the internet doesn't exactly help them realize that they have a problem. [The] problem is that the research chemical market is a big candy store where everybody is invited to have a go at everything. And some of the stuff offered is really detrimental but too alluring to be left alone.'

As drug users in Europe were gorging themselves on mephedrone in 2009, the original research chemical scene was bracing itself for a legal clampdown, unhappy at the attention it was now receiving. Then late that year a disturbing series of overdoses shocked even seasoned observers. Mislabelling is perhaps the gravest danger facing research chemical users. If you order one drug with an active dose of 10 mg and receive by accident another drug that is ten times more potent, there's every chance you will overdose or die.

On 3 October 2009, a twenty-two-year-old Danish man, Dannie Haupt Hansen, took an 18 mg dose of what he thought was 2C-B-FLY, a potent psychedelic phenethylamine that is an analogue of Shulgin's 2C-B, which he had ordered from a Chinese lab. Before he tried the drug, Haupt had also been selling it from his website, haupt-rc.com, to customers all over the world. Haupt died of a heart attack, for the drug, in common with many psychedelics and stimulants of this class, is a powerful vasoconstrictor – his heart gave out as it worked to pull the blood through his tightening arteries. But he took the 'right' amount – he had researched the drug and its dose online. What had gone wrong?

As Haupt was overdosing and dying, in California eighteen-year-old Brian Sullivan was nodding his head to the music, tripping with his brother John and his brother's girlfriend. They too had taken the Danish batch of the drug, ordered online and sent under plain mail to the West Coast of the US in a few days. They too were overdosing. Sullivan died; his brother and girlfriend convulsed, but escaped with their lives. And they too had only taken a small amount – far below the safety levels they had researched online in trusted forums and in chemistry journals.

On 5 October, Erowid and Bluelight posted an alert online warning other users that a potentially lethal batch of 2C-B-FLY was circulating worldwide.

In Barcelona that same day at 5 p.m., a young man was staring at his home-made pizza. The topping of goat's cheese and tomato suddenly looked like pools of pus and blood. A few hours earlier, he had been weighing some 2C-B-FLY and some other research chemicals he had bought, including some synthetic marijuana compounds and an MDMA analogue. He scooped up a tiny crumb of a white powder that had spilt on the table and wiped it onto his inner lip. He lay down to relax, closed his eyes and put on the Beatles LP *Rubber Soul*, but was quickly assailed by powerful hallucinations. He didn't know it, but he too was overdosing. At 5.24 p.m., afraid and overheating, he logged on and fired up a Google search. He hit Erowid and found an emergency posting about a spate of overdoses caused by a dangerous batch of 2C-B-FLY.

He described his experiences in an Erowid report a week later under the name Joan Miro. 'My heart dropped when I saw the photo of the same exact bag that was sitting on my desk. I became angry, agitated.'[1]

He logged on to Bluelight and begged for help. He was told he had to get to the nearest hospital immediately. His life was saved by the bulletin board – whose posters told him he had a six-hour delay before things worsened dramatically – and the

Catalan hospital's staff, who tranquillized him and monitored his heart rate and temperature.

Miro said he doesn't remember much of the next few hours other than fear and panic and guilt over impending and self-induced death, all amplified and distorted by the powerfully psychedelic drug, in a chaotic emergency ward filled with screams. 'I thought a lot about how this would be such a horrible way to go. Especially for my family and friends. What would it do to them for me to leave them like that? Over something so stupid [and] preventable and – especially – without telling them goodbye and that I loved them. I was determined to walk out of that hospital,' he wrote. He survived.

A poster named Voltech posted a message on Bluelight on 16 October containing the results of lab tests on the compound that had killed and sickened the users so dreadfully. It turned out that the Chinese laboratory had accidentally sent a far more potent drug, bromo–dragonFLY, in place of the more benign 2C–B–FLY. Bromo-dragonFLY had been synthesized originally in the lab of David Nichols in 1998, and was designed to map the topography of the brain's serotonin receptors. It was not designed for human consumption, but rather for running lab tests using brain tissue. (Even in the petri dish or under the microscope, brain tissue can react to chemical agents allowing data about their action to be gathered, and valuable new medicines to be invented.)

These superficially very similar images (opposite) of two vastly different drugs show how even the tiniest of differences can change a chemical radically and dangerously, reducing its active dose dramatically, thanks to its greater receptor-binding qualities. Bromo-dragonFLY, so named because images of its winged molecular structure resemble an insect in flight, is active from just 200 µg – one-fifth of one milligram. Haupt, then, had taken ninety times the active dose.

*2C-B-FLY, active at 18 mg*

*Bromo-dragonFLY, active at 200 µg*

Nichols' results were published in a 1998 research paper in the *Journal of Medicinal Chemistry*, entitled 'A novel (benzodifuranyl) aminoalkane with extremely potent activity at the 5-HT2A [serotonin] receptor'.[2] In common with all formal, peer-reviewed chemistry texts, it would have included the method of synthesis. The Chinese chemists possibly used this recipe to synthesize the wrong compound, and the young Danish vendor and his customers had failed to carry out adequate safety checks. These might be simple reagent tests, or more complex third-party lab tests such as nuclear magnetic resonance analysis, mass spectrometry, or high-pressure liquid chromatography tests. It may be that the vendor ordered the wrong drug, and the Chinese lab actually got it right. The truth will never be known. The drug that was supposed to have been sent, 2C-B-FLY, was legal at the time in Denmark, and it still is.

At the time, bromo-dragonFLY was also legal in each of the countries involved in the transaction – China, the UK, the US and Denmark – or at least unscheduled, as long as it was not sold for human consumption. Predicting the ingestion of every possible psychoactive analogue or chemical is impossible, and so the law must at times be reactive. When innovation is running at a drug a week in the EU, the rulebook has to be ripped up and a new, more intelligent approach devised.

Only Sweden, Denmark, Norway, Romania, Australia and Finland have now banned bromo-dragonFLY. (In Finland, in one of the oddest stories to emerge from the research chemical scene, it was banned after the drug was used in a bizarre murder by a psychopath who stabbed her victim with a syringe full of it.)

These stories explain the current super-stringency of customs and border control in Scandinavia. But they pose more questions than they answer. If a classical view of drug abuse correlates it with indicators of social deprivation such as unemployment and poor educational opportunities, why, then, are the rich and prosperous Scandinavian countries – above all, Sweden – so over-represented in the online designer drug world? Why have there been far more overdoses and deaths due to the more novel psychoactives in Sweden than in neighbouring countries, such as Germany? The answer might be found in the country's drug laws, which are today the most stringent in Europe, and have been for decades. In Sweden potential employers can demand your criminal record before interview, and fifty per cent of them choose to do so. To have a drugs conviction in Sweden is to be an instant member of the underclass, with education and employment opportunities denied outright for many. The country's extremely punitive drug laws are rooted in the work of medic-turned-drug-campaigner Nils Bejerot, a forceful hard-liner who was a lone voice in the 1960s, when Sweden was more liberal on drugs, but whose influence endures twenty-four years after his death. In a report for the Swedish Carnegie Institute, drug policy analyst Jonas Hartelius noted:

Bejerot showed that while the allure of drugs is biological, the level of drug use in a community, in a nation, and in the world as a whole, is largely determined not by brain biology, but by the social reactions to drugs. Tolerate or, even worse, encourage drug use and drug use explodes in a deadly, self-propelling behavioural epidemic. Identify drug users, reject their drug use, and insist on enforced abstinence, and the drug epidemic is quelled. Drug policy really is, Bejerot argued, that simple.[3]

In the Swedish context, if you accept the simple but, to some people, unpalatable truth that sometimes humans want to alter their state of mind, the search for legal alternatives becomes more understandable. Bejerot's view of 'community' and 'social reaction' were formed in a pre-web age, and any notion of enforcement must acknowledge that dealers and users can now act invisibly. Dannie Haupt's drugs, which were legal, were sent to him and his customers worldwide in the post.

Bejerot's intentions were doubtless good. But to persist with his drugs policy in this very different, globalized internet age, is, in my view, ill-considered and ignorant of the basic realities of twenty-first-century life.

In July 2012 Swedish coroners reported that 5-IT, yet another new research chemical, a stimulant that had escaped British legislation, had been found in fourteen post-mortems.[4] Information was patchy at the time of reporting, but the drug disappeared from research chemical vendors' sites within days of the news emerging. The usual cause of death in these cases is from serotonin syndrome, where the brain becomes overloaded with an excess of the neurotransmitter and the unfortunate victims overheat, then convulse until they die.

At least Sweden is consistent in its hardline approach to intoxication: alcohol over 3.5 per cent proof can only be bought in government-approved shops, known as *Systembolaget*, or the System Company. On the other hand, headshops and websites there are selling potent research chemicals with impunity.

Another case of mislabelled research chemicals was reported online in August 2010. Agnetha, a Bluelight poster, was an expert user, and had spent years refining her knowledge of her sources, the drugs and their effects. She knew the dosage of each of the chemicals she took, and approached them with an almost scholastic vigour; she was a curiously puritanical hedonist. She had read all of Shulgin's work, and had methodically set about trying dozens of the compounds in the books. But when she found herself hallucinating on a stimulant, she knew something was wrong.

The scene's biggest supplier at that time, with fulfilment centres in Taiwan, had sent her a package containing 2C-P – a Shulgin psychedelic active at 8–12 mg. But the bag was mis-labelled as buphedrone, a mephedrone variant with a standard dose of around 80–100 mg. Agnetha overdosed twice in a week. She described the experience in a post on Bluelight:

> The chemical took five hours to come up. It started with sweating profusely and uncontrollable muscle spasms and got worse from there. The trip lasted over 30 hours . . . The night was chaotic and apocalyptic, no sleep, heavily disorientated. No music, no light, the visual and auditory distortions were too heavy anyway to comprehend any external input. Psychotic and delirious would be an accurate descriptions of my mindset in those hours.[5]

A week later, she overdosed again, having assumed that the error the first time was hers – she thought that she had confused two of the many bags in her drug collection. But she had not got her chemicals mixed up; the vendor had. She wrote:

> The second incident was one week later when I finally

came around to test the stimulant as I originally intended. There was no mistake this time. The RC vendor had mislabelled the bags. He sent me a very potent unknown psychedelic – most probably of the 2C family – and packaged it with the label 'Buphedrone'. I overdosed again on 80 mg. Very frightening prospect. It was one of the more respected vendors. I was reminded of the bromo-dragonFLY fuck-up that had cost several people's lives a while back.

I loved the moment the sun came up. I spent dawn naked on my terrace. The world around me looked like a churning and swirling Van Gogh painting – only much sharper, more precise. A naked animal, poisoned and exhausted . . . I knew I would survive this, there was a way out . . . I called in sick . . . but had to type some emails, make a couple of phone calls and review some texts. In hindsight I actually produced high-quality work that day, enhanced by crystal clear, super-precise psychedelic thinking.

Agnetha survived, no thanks to the vendor. These kinds of labelling mix-ups are mercifully rare in the online drugs market, but the dangers are real and the consequences can be deadly, no matter how statistically improbable.

There were nine other users poisoned in this incident. Agnetha traced most of them. 'The owner of the site didn't mention the incident with so much as a single word on his site in the eight weeks or so it endured after my accident. He somehow found time to announce a couple of very appetizing sales within that timespan, though,' she told me by email.

Disturbingly, as we move out of the post-mephedrone era many more powerful drugs are available now to many more people than ever before, and the culture and practice of buying drugs online in this way is becoming more prevalent. As more drugs appeared in the UK after the mephedrone ban in 2010, the number of deaths also increased. Drugs such as methoxetamine gained great popularity worldwide in 2010 and onwards. Some found it numbed chronic pain or helped lessen their neuroses, thousands more enjoyed the semi-alien buzz the drug gave and ramped up their doses until they slipped into a space outside time, geography and human interaction, lying inert on their sofas and floors, but travelling thousands of miles inside their minds. Some users, perhaps not knowing anything more than that it was a legal white powder that got you high, were merely overwhelmed by the drug; the unluckiest died.

Reports of methoxetamine addictions and patterns of problematic use surfaced within months of the chemical's release, with users bingeing until they were delusional or psychotic. The first methoxetamine death was in December 2010, when a user in Sweden intravenously injected the drug along with a massive dose of the Nichols-designed serotonin agonist, MDAI. Intravenous drug use is ill-advised even for chemicals such as heroin and cocaine, drugs that have been used in this way for decades, albeit perilously. Injecting drugs means they are not metabolized by the liver, or digested by the stomach, bringing an extreme rush into the bloodstream, and the brain, seconds after the injection. This intense hit is sought by users who either want to conserve their drugs – making them go further, since less is required – or by those hellbent on the most extreme high at any cost. To use research chemicals this way is extraordinarily dangerous, and indicative of a serious drug problem. The death was reported on the Swedish drugs forum, Flashback, by a poster named Miss Tranquil. 'His heart started beating about a million beats per second and then *bang*. Dead. On my couch. So take it reeeeal easy.'[6]

Fred, an author and musician in London, took methoxetamine on his forty-fifth birthday in a London pub in early 2012. He was unprepared for what happened next, he told me by email:

It was a Sunday afternoon pub birthday. All very civilized and middle-aged. As evening turned to night and the kids went home with the babysitter, someone came back from the bar with a round of shots, which was only thirty seconds behind the round of shots we'd just sunk. With this sudden bolt of hedonism, my thoughts turned to a line of coke, or anything else that might put an adventurous shape on the night. Most people there were no strangers to class As and a quizzical muttering went round the room, only to come up flat. 'I've got this,' said a mate unenthusiastically, passing me a wrap. 'We got it off the net, and we did it in Berlin. It's horrible but you might like it.'

Hardly a recommendation. But I'm a curious type. I've taken my fair share of pills and powders in two decades of drug-taking, with Ecstasy, cocaine and ketamine regularly on the menu, as well as the occasional acid or mushroom trip. I even did DMT once, and rattled around Burning Man 2005 guzzling the postcode drugs (2C-B, 2C-E and their chums). I asked my mate more. Turns out it was something synthetic. 'One of those new ones, off the net,' he said. The important thing was to only do a tiny amount. He stressed this: 'Just half what you'd do if it was K.' I knew about the new synthetics and I knew dosage was a serious thing: overdo it even a little and you'd be cabbaged or worse.

All I wanted was some casual dislocation, so as I went in the cubicle foremost on my mind was to take a tiny amount, to be really careful. But as the door closed, the alcohol ambushed my brain, rational thought evaporated and I fell into autopilot – just a drunk in a pub doing what I'd done so many times before: taking a bump in a toilet. The only notion in my head was the automatic 'Do what fits on a door key . . . twice'.

Had it been one of the old familiars, that animal rule of thumb would have worked fine. I was ready for the bendiness of ketamine, followed by its reasonably swift return to normality, or the sharpening effects of coke. But instead I had overdosed on methoxetamine. Everything in my experience said my little hit should have just dented the edges of reality for a while. Nothing I couldn't handle. My instincts told me the night would continue fairly unchanged. I'd carry on chatting to my friends, just with a slight twist.

Instead, within minutes I was slumped on a sofa unable to move or talk. I fought it valiantly, but I was poleaxed. Most of the next three hours is gone, the only real memory is the awful reality of puking in public. Not exactly the urbane sophisticated drug-taker I'd been for the last decade. As for the drug itself, it felt like a really pointless version of ketamine: no psychedelic effects, no pleasant slide into rubbery nonsense, just a sudden drop off the cliff of wrongness. The alcohol mix no doubt made it doubly unpleasant. Hours afterwards it still felt like the switch was in the wrong position. I was doped and wired at the same time; I couldn't even sleep it off.

For me, that was the day it changed. Never again will I bumble my way into an unfamiliar powder trusting to instinct and experience. The care you need to take with these new drugs isn't compatible with a feral night out. My generation is ingrained with a set of rules for safe and sociable drug-taking, rules that have served us well for a long time. Well, the old rules no longer apply.

This is a message that bears repeating: the rules of chemical engagement have changed.

On 30 January 2012, two men under the influence of methoxetamine died in Canterbury. Popular and talented busker Daniel Lloyd, aged twenty-five, and his beat-boxing friend Hugo Wenn, aged seventeen, were both found drowned in Reed Pond, near an army barracks in the town. Hugo's mother Fiona told the *Kentish Gazette* about her son's rural upbringing, and how she had hoped this might have offered her son protection from the dangers of drugs: 'We often talked to all our children very bluntly about the dangers of drug-taking but Hugo grew up in the village in the countryside and none of his friends were into that sort of scene. It was so far removed from our lives. There was never any sign or suggestion in the past that Hugo was taking drugs. It just wasn't him.'[7]

In February 2012 a fifty-nine-year-old woman and a thirty-two-year-old man were found dead at their homes in Leicestershire, after taking a methoxetamine overdose.[8] That same month Andrew Cooke, a twenty-nine-year-old drummer from Crystal Palace, went missing in east London and it is thought he may have consumed methoxetamine – knowingly or unknowingly – in the hours before his disappearance. 'He was last seen between 3–4 pm Sunday afternoon (February 12) at a free/squat party located on 1 Lea Valley Road near Chingford,' his friends wrote on Facebook. His body was found in a nearby canal on 14 March.[9]

After the government ban was announced in late March that year, Sally Bercow, the wife of John Bercow, speaker of the House of Commons, tweeted, 'Mexxy is a legal high that is, er, no longer legal. And now we've all heard of it, demand will rocket.' And with that, the research chemical scene was at the heart of British political life, reported upon in Middle England's tabloid of choice, the *Daily Mail*, and tweeted about by the wife of the Speaker of the House of Commons.[10] In every major city in the UK and many small towns there were people buying new, untested and powerfully psychoactive chemicals marketed as legal highs, with no indication of what the drug was, what it did, or how it should be taken.

The most obvious response to these tales is to preach a message of personal responsibility, but when the new drugs are this powerful it's highly unlikely that people will be able to dose correctly, even if the fault ultimately lies with the user rather than the drug. Many of the new families of drugs are too potent to use in any casual setting by either experienced or uninformed users. They are not party drugs.

We need a more nuanced, insightful approach than banning new drugs as they appear, as simply expecting people to stop buying, selling and using them is unrealistic. The knowledge exists, the drugs exist and the market exists – what is lacking is education and a new legislative approach.

Dr Adam Winstock agrees that wider society is not ready for the greater availability and more novel research chemicals that have followed since mephedrone. He told me, 'The internet parachuted these new drugs into user groups, and that meant there was no way for people to accurately discuss correct and safer use around things like dosage and onset of action. So with methoxetamine, the information that it was active at just 10 mg, with a slower onset of action than ketamine was lost, and people redosed dangerously. Government action and blurring of legislation means we can't tell people that effectively. It should be on the label!'

The most worrying aspect of the recent new growth of the research chemical scene into the mainstream is that many new users are not observing the most basic principles of harm reduction. Expert users have long stressed the importance of knowing and trusting your source, but users are now simply buying from the first, or cheapest, or most convenient source. The only way to know for certain if the correct substance has been sent is to have it tested via expensive chemical means; vital nonetheless, as even an allergy test of 1 mg might be an overdose.

There is a sense of unregulated, late-stage capitalist anarchy in the online research chemical scene at this point, in 2013. For people who had been watching the story develop, the emergence in 2009–10 of public forums with site sponsors using banner ads to offer cut-price research chemicals of every hue was a death knell. In the early days of the online designer drug scene people were barred from forums or listservs (email subscription lists) for asking for sources. Twenty years later, there were links alongside and below forum posts to vendors of chemicals that hadn't been tasted by any human beings on earth at all. In the past, conversations about newly synthesized chemicals were carried out in what were essentially private members' clubs, digital speakeasies known only to a few old trippers and radicals. Since 2008 or 2009, there has been an endless real-time stream of conversation, all public, all unmediated, about where to find and buy and sell drugs that did not even exist five years ago.

'The caution and concern back in the early days was that sharing sources openly and discussing the chemicals explicitly might prompt a bust or, worse, a chemical being specifically named and banned in the US, as plenty were after Operation Web Tryp,' one user told me. 'Most of all, the *omertà* on sourcing was a bulwark against stupidity, a safeguard in a world where no rules applied. The thinking was that if you didn't put the legwork in, you probably weren't clued up enough to use these drugs. After mephedrone, it just went silly, mental, there was so much money to be made. The main reason people didn't give sources back in the day was because a good number of these drugs could kill you – even if you were actually sent the right compound. Who wants that responsibility?'

Mislabelling, then, is not the only danger. Ignorance kills just as fast. There has been a spate of deaths among teenage users in the US in recent years. They may be choosing research chemicals because they have less access to traditional drugs, and they may be more foolhardy, and less knowledgeable about the effects of the drugs. In the small American town of Blaine, Minneapolis, on 17 March 2011 a group of high school students shared the drug 2C-E at a spring break party. The night ended in a mass poisoning and a fatal overdose from this potent, Shulgin-devised psychedelic. Various drug forum users suspected that a mislabelling had occurred as in the Haupt case, or as in Agnetha's case, and there was a palpable sense of fear. However, forensic tests showed that the drug taken was indeed 2C-E – it had just been dosed wrongly and dangerously.

Trevor Robinson-Davis, the nineteen-year old father of a five-month-old son, was taken to hospital after snorting a large line of the drug. He became instantly agitated and died of a heart attack. The teenagers had not used a scale to weigh the drug out and had instead 'eyeballed' it – judging a suitable dose by looking at it – with fatal consequences. 2C-E is active at around 8 mg, and has an extraordinarily high dose-response curve, meaning that 18 mg of the drug will hit you far more strongly than even 14 mg. The young people who survived reported delirium, hallucinations, paranoia, auditory distortions and overheating. Timothy Lamere, a twenty-two-year-old, was charged with third-degree murder for supplying the 2C-E and was jailed in 2012 for nine years and nine months – the longest available term under local sentencing guidelines.

Lamere told the court that he bought the drug online and believed it be legal; at the time, it was not specifically scheduled in the US, and its status as an analogue had never been debated in court. In an unusual move, federal prosecutors intervened in the state case, threatening to escalate the charges if the court did not hand out the maximum sentence possible to Lamere, who had previously been admitted to a psychiatric ward for complications with bipolar disorder.

The town's local newspaper, the *Star Tribune*, hosted videos of the survivors of the night. Katrina Loomis told journalist Pam Louwagie: 'I think about it all the time, every day. Constantly probably. I will never touch another drug. If we'd just been smarter and thought about what we were doing before we did it, we would still have our friend here. And Timmy was our friend. So we lost two friends that day.'[11]

Elsewhere in the US, the research chemical scene has spilled into headshops from its online roots, just as it has in the UK. 'Bath salts' are the American iteration of the 'plant food' craze seen during the UK mephedrone craze, whereby vendors dodged the law by marketing new and potent drugs with a nod and a wink and a fake label. A thriving market for bath salts and fake pot started up in the US in 2010, and until 2011 both were sold legally by some tobacconists and other shops. The products contained a very wide range of substances. Bath salts contained stimulants such as MDPV or 2-DPMP, flephedrone or mephedrone, none of which was scheduled in the US at that time. The shops also sold 'incense blends' that actually contained the then-legal JWH-series of cannabinoid receptor agonists.

Some of the brand names for bath salts – surely the greatest misnomer ever for drugs that caused palpitations rather than relaxation – included Bliss, Blizzard, Blue Silk, Charge+, Hurricane Charlie, Ivory Snow, Ivory Wave, Ocean Burst, Pure Ivory, Purple Wave, Red Dove, Snow Leopard, Star Dust, Vanilla Sky, White Dove, White Knight, White Rush and White Lightning.

Branded synthetic marijuana products also started to sell to millions of users in the US, via gas stations, convenience stores and skater shops, with names like Blaze, Dream, Aroma, Mr Smiley, Red X Dawn, Kush, K2 and Abama (perhaps a pun on the famously pot-puffing president's name); they cost ten to twenty-five dollars a bag and contained JWH-series drugs, just as the Spice branded legal highs had in the EU.

By mid-2012 in the US, media reports about users behaving in violent or bizarre ways after taking bath salts took on a surreal, filmic quality. Perhaps the most famous case reportedly involving bath salts was that of thirty-one-year-old Rudy Eugene, who will be grimly remembered as the man killed by police after biting off the face of fellow homeless man, sixty-five-year-old Ronald Poppo, in Miami in May 2012. Stories flashed around the world instantly, with police who were not present at the scene of the crime saying it was likely that the attacker was under the influence of a bath salts type drug. Eugene was impervious to gunshot, police said, and took several bullets from marksmen before he died.

The American Center for Disease Control and Prevention lost its mind that week too, and responded, in all apparent seriousness, to online rumours of a zombie apocalypse following a rash of other reports involving cannibalism in the US and elsewhere. 'CDC does not know of a virus or condition that would reanimate the dead (or one that would present zombie-like symptoms),' agency spokesperson David Daigle told *The Huffington Post*.[12]

There is no doubt that Rudy Eugene was mentally ill and it is possible that he took drugs on the day of the attack that made his condition worse. But at the time of initial reports, there was no confirmed evidence that Eugene had actually taken bath salts. Some undigested and as-yet unidentified tablets were found in his stomach, but the branded legal highs sold in the US are typically powdered and are snorted. No human flesh was found in his stomach, meaning initial reports of cannibalism were also inaccurate. Local TV station CBS4 blamed an 'LSD-type drug' for the man's attack. Leaving aside the fact that no toxicology reports were available at the time of that report, and that no bath salts type drugs are anything remotely like LSD, the source for these allegations was flimsy at best – neither the doctor nor the policeman quoted in the early stories had first-hand knowledge of the case, reported Reuters' Jack Shafer.[13]

It is complex enough untangling the facts in news stories that involve novel psychoactive substances without media hype confusing the picture so completely that it seems almost wilful.

Each generation has its drugs moral panic, whether it comes in the guise of LSD users jumping from buildings in the 1960s, PCP-crazies in the 1980s, superhuman crackheads in the 1990s, or Meow-frenzied and entirely fictional schoolkids taunting their teachers with their bags of legal highs in the early twenty-first century. As Alasdair Forsyth, of Glasgow Caledonian University's Institute for Society and Social Justice Research, told me, 'There was a cartoon in *Punch* a few years back depicting two farmers looking at a huge scarecrow with the caption: "To have any effect I find I have to make it more scary every year".'

Any retelling of the Rudy Eugene story is incomplete without a wider analysis of the sociocultural and economic climate in which it occurred. It is certainly not as grimly compelling as the possible news of a zombie cannibal apocalypse, but no mention was made of the fact that Florida has the second-worst funding of mental health services in the US. There are 325,000 adults with severe and persistent mental health problems in Florida, and only forty-two per cent of them receive state support, found the Florida Center for Fiscal and Economic Policy in a 2009 report. That means over 190,000 seriously mentally ill people do not receive the help they need.[14] In March 2011, Senator Joe Negron of the Appropriations Committee proposed a further two-thirds cut to the mental health budget in the state. Banning a drug allows politicians to appear in control, but the problems that led Rudy Eugene to almost kill Ronald Poppo can't be solved that easily.

Edward Huntingdon Williams MD wrote in the *New York Times* of 8 February 1914 about the dangerous new cocaine craze sweeping through the 'shiftless' black working-class communities of the south.[15] The piece displays all the casual racism so common at the time, but leaving aside that easy target, it's interesting to note how the central meme – that of a terrifyingly violent black man made superhuman by drugs and impervious to gunshot – prevails a century later. Bigger guns were needed to control these negroes, it was decided in 1914. While the so-called Miami zombie cannibal case inspires both pity and a sheer visceral terror, the reports conflated ethnicity, drugs and violence just as other racists had done a decade before.

Bath salts were involved in other bizarre news stories in the US in 2012. The flesh-eating virus had taken hold more quietly earlier that year when the American medical journal *Orthopedics* reported that a woman had suffered a bout of necrotizing fasciitis contracted after injecting a dose of bath salts into her arm. Medics reported that the flesh-eating disease crept through her body even as they watched, and moved so rapidly that they had to amputate her arm, collarbone and shoulder and then perform a radical mastectomy.[16] But the disease had nothing to do with the drug in question, more the route of administration – an intramuscular or subcutaneous injection – and the patient's existing health conditions. Most casual drug users do not inject themselves, and it is a fair assumption that habitual injecting drug users have worse health and correspondingly weaker immune systems than other drug takers.

A measure of the drugs' popularity and the dangers associated with their use was seen when the American Association of Poison Control Centers reported that it took almost 6,000 calls related to bath salts, and 7,000 related to fake pot in 2011. In 2009, there had been none at all. Louisiana Poison Control Center Director Dr Mark Ryan told ABC news, 'It doesn't matter which socioeconomic strata you're from, we're seeing these drugs being used across the board – all ages, all economic groups. We've had some people show up who are complaining of chest pains so severe that they think they're having a heart attack. They think they're dying . . . They have extreme paranoia. They're having hallucinations. They see things, they hear things, monsters, demons, aliens.'[17]

In the southern state of Alabama in May 2012, thirty young

people were admitted to hospital wards with kidney failure after smoking the herbal blends that had been sprayed by accident with a pesticide as well as the synthetic cannabinoids. Some of those afflicted will be on dialysis for life, doctors said.

The sheer range of branded legal highs in the US shows the popularity of the drugs, and the prevalence of their use, while the profitability of the drugs when sold in branded sachets is extraordinary. One single gram of MDPV bought in bulk for three dollars in Shanghai can be made into forty packets of a branded high sold at US$25 in American convenience stores – a profit of US$997 before packaging and distribution. The profit margins for the drugs in the JWH-series and other cannabinoid chemicals is just as high.

It is clear that the US has taken to bath salts, synthetic pot, and other research chemicals. Why it has done so is a thornier and much harder issue to identify. The country is the world's number one consumer of regular drugs, with the 2010 National Survey on Drug Use and Health revealing that twenty-two million citizens – nine per cent of the country – use illegal drugs. In those aged eighteen to twenty-five, that proportion more than doubles, to 21.5 per cent. It is noteworthy that this report – published in 2011 – contains no mention of the new drugs that have become so widely used in the US, although the internet and TV channels are alive with social chatter and news reports of their use.

Perhaps a driver in this market is the far wider use of drug testing in American firms, and the use by some American parents of testing kits on their children. College sports scholarships demand regular drug testing; fail and you're not just off the team, you've lost your scholarship and you're out of college. Many of these new chemicals will return a negative test result.

Consider, too, that in 2011 three dozen states proposed drug testing for people receiving welfare, job training, food stamps, public housing and unemployment assistance. The state of

Georgia was the most recent to pass the law in April 2012 and now compels 'some' benefits recipients – it's not clarified which criteria are used – to submit to drug testing before essential benefits are paid. Not satisfied with targeting the welfare payments of some of the poorest and most needy in society – penniless addicts and drug users – the state also demands that those targeted pay seventeen dollars to urinate into a testing vessel that will decide their fate. The *New York Times* reported that in Florida 'people receiving cash assistance through welfare have had to pay for their own drug tests since July, and enrolment has shrunk to its lowest levels since the start of the recession.'[18]

But when drugs active at just a few milligrams are sold to anyone with thirty dollars to spare, the blame lies not only on the labs in China who make them knowing full well that they will be sold as drugs, and on the shops and websites in the US that sell the drugs, but most of all on a legal system that has made the sale of these compounds profitable and their use attractive. The reason these drugs are causing deaths, overdoses and delusion is because they are being sold under false pretences as cheap and legal alternatives to drugs like MDMA or cocaine or marijuana.

While official responses are failing to effectively address the situation, grassroots voluntary organizations are taking direct action to ensure the health of festival-goers. The Bunk Police is a US-based group that produces test kits that use a simple set of reagents to identify – or at least test for the presence and absence of – certain chemicals. Small samples of drugs are placed in a small plastic tube and each reagent is added to the mix. The colour changes depending on what chemical is present, and this is then compared to a printed chart. Bunk Police also conducts more complex lab tests on substances that have hospitalized users. The group checks drug quality at raves and festivals, and also distributes test kits to drug users so they can perform these basic tests themselves.

'We started in June 2011,' the group's spokesman, who remains anonymous, told me by email. He went on:

> When we were doing live testing it would depend on the venue and amount of traffic we had. Now that we distribute test kits, the number has increased dramatically but still depends on those factors. We've distributed well over 1,000 kits at a single four-day event. We operate at music festivals and other rave type events that offer camping [operating among the tents and speaking privately with users]. We also distribute portable test kits that can be concealed and used in a crowd at smaller events.
>
> Our objective is to put an end to the dishonesty that goes on in the black market. In most cases, those who choose to take illegal substances have no way to tell if what they are taking is real. Some of the substituted chemicals can be much more dangerous than what the user intended to take. We find poor-quality drugs more often than not, but it really depends on the event – some are much worse than others.

Individuals in the US are now buying research chemicals by the kilo from China and substituting them for regular drugs more often than they are actually selling the regular drug itself, he explained:

> The best example is that synthetic cathinones and other stimulants are sold as "molly" (American slang for powdered or crystal MDMA) more often than MDMA. There are only a handful of substances being offered (LSD, MDMA, cocaine, mescaline, etc.) but in fact these substances could be any of over fifty research chemicals. The dangers with RCs are in the unknowns. There is

very little formal research on these substances, which means that there could be any number of hidden dangers associated with them. The dangers associated with taking an unknown substance or a mixture of unknown substances is also very substantial.

In this book's final chapter I will attempt to untangle the legal and social complexities the increase in new drugs have brought about, and suggest ways that might reduce the likelihood of further tragedies in the future.

In this celebrity-obsessed age, a tipping point whereby these drugs entered into the national consciousness, at least in the US, occurred when actor Demi Moore's friends' called 911 as she suffered a terrifying reaction to a synthetic cannabis compound she had taken. In a call posted to YouTube in December 2011, Moore's friend says to the operator as the star suffers fits next door, 'It's not marijuana but it's similar to – it's similar to incense. And she seems to be having convulsions of some sort.'[19]

That Hollywood stars, who have ready access to the purest and most exclusive drugs on the market, are choking down damiana and dried sage sprayed with John William Huffman's cannabinoids demonstrates beyond doubt that the research chemical market has penetrated areas of society unthinkable a decade ago. While it is true that the new drugs scene has in many cases been caused by a scarcity of a preferred product, it is also true that for many drug users, a new kick is always welcome.

And while many drug users are avoiding the law by making or taking drugs that lie outside international control, a high-tech anarchist cadre has built an online market in traditional narcotics that may very well be untouchable, and which represents a major new battlefield in the war on drugs.

# 9

## *Your Crack's in the Post*

The first time you see the Silk Road website there's a creaking disconnect between your eyes and all the evidence they deliver, and your preconceptions up to that point. There's a strange smile, mixing recognition, revelation and confusion, playing on your lips. It all looks familiar to anyone who regularly shops online, but is in some way uncannily different. It can't be real, can it? Yes, it is. You can buy any drug you want right now on the web. Every drug you can think of, and a dizzying few dozen more, are on open sale on the site, from old-style illegal Class A drugs such as crack, heroin and LSD, to research chemicals of every hue, including Shulgin's creations, Nichols' work, Karl's ketamine variants and Kinetic's mephedrone. If there's a drug missing that you really want, you can always ask for it to be offered, imported or synthesized. And if you're a chemist yourself, there are syntheses and precursors for sale too.

Rather than waiting for the drug laws to change, activists, dealers and users have declared an independent state online where all commerce, within certain boundaries, is permitted by the site's mysterious owner. Its guide to vendors is pretty laissez-faire: 'Do not list anything who's [sic] purpose is to harm or defraud, such as stolen items or info, stolen credit cards, counterfeit currency, personal info, assassinations, and weapons of any kind. Do not list anything related to pedophilia [sic].'

The site has Norwegians selling Cambodian mushrooms, Canadians selling Afghan heroin, and Brits selling concentrated cannabis tinctures from ancient Nepalese cannabis landraces grown under artificial sunlight in lofts that may well be in Basildon. Appropriately for a site named after a trade route that first brought these drugs to the West, there are also opiates, including opium, prescription morphine, and white and brown heroin from Afghanistan. Most of the products are illegal, but whether you want a quarter gram of heroin or a gram of glittering Peruvian *escama de pescado* cocaine, you're in the right place, and there's not a great deal the police or customs can do to stop you.

The Silk Road is a cyberpunk dreamland – except it's happening today, in dozens of countries, not in some dystopian future in a William Gibson novel. For all that, it's a website like any other, if web design skills were locked down in the ancient pre-Google, pre-Amazon days. You almost expect each page to be soundtracked by the screaming of a modem while the bits crackle slowly down the phone lines. But instead of books or household goods, there, in the most stark and simple language possible, advertisers lay out the drugs they offer and their prices. In a neat left-hand navigation bar there's a list of different categories. Psychedelics are well represented, along with research chemicals and standard options such as LSD and an abundance of mushrooms. There are 2C-B, 2C-I and a few other Shulgin-created delicacies for the chemical cognoscenti. There's DMT – the drug William Burroughs travelled months in the Amazon to find. One vendor, Seakong, has a gram of the more hallucinogenic cousin of MDMA, MDA, for sale, synthesized, he says, by an aspiring chemist friend. At just ฿12, it's an absolute steal for such a rarely seen drug. The ฿ symbol stands for bitcoin, the mysterious currency whose use is compulsory in this online market, of which more later.

There are tranquillizers, such as Valium, lots of crystal MDMA and Ecstasy pills, dissociatives such as ketamine, and stimulants, including crack, ice and other amphetamines. There's high-grade kush marijuana, with enthusiastic recommendations from satisfied customers for one particular vendor, detailing how he vacuum-sealed and wrapped and triple-packed the highly fragrant goods into an envelope small enough to be posted through most standard letterboxes, negating the need to sign for the packets – or for the raising of any red flags at customs. That the strain is one of the world's oldest and earliest genetic examples of the plant, brought to Europe and thence to the US along traditional trading routes, is an irony probably not lost on the Silk Road's intelligently combative and articulate owner, who operates under the pseudonym Dread Pirate Roberts.

Buying is a simple matter of adding the goods to your shopping cart, and paying for them. The money is held in an escrow account hosted at the site, and although you have to supply a delivery address, this can be encrypted, and is deleted as soon as you have received the goods. The site also gives detailed information on how to receive packages safely:

Use a different, unrelated address than the one where your item will be kept, such as a friend's house or P.O. box. Once the item arrives, transport it discreetly to its final destination. Avoid abandoned buildings or any place where it would be suspicious to have mail delivered. Do not sign for your package. If you are expecting a package from us, do not answer the door for the postman, let him [deliver it] and then transport it as described above. Do not use your real name. This tactic doesn't work in some places because deliveries won't be made to names not registered with the address. If you think this is a problem, send yourself a test letter with the fake name and see if it arrives. If you follow these guidelines, your chances of being detected are minimal. In the event that you are detected, deny requesting the package. Anyone can send anyone else anything in the mail.

Initially the Silk Road also had a weapons trading area, but many users were uneasy about the influx of arms dealers and a new subsite, the Armory, was launched in February 2012. It was closed in August 2012 due to a lack of interest.

The Silk Road's turnover reached US$22 million a year within its first year of operation, according to security researcher Nicolas Christin, who scraped the site's data by deploying software agents under multiple user accounts that recorded customer activity via the public feedback system and showed how many transactions had taken place. He crunched that data in the middle of 2012 to calculate the market's size.[1] The site's owners take a commission on each sale of around six per cent – or US$143,000 per month at current rates.

The Silk Road has a very busy forum area, too, with over 100,000 posts, 9,000 topics and 11,000 users in the bustling community pages. The conversations there weave around the site's holy trinity: drugs, smuggling and cryptography. There was once even a post purportedly by a Canadian postal official who claimed to have become addicted to opiates following an accident, in which he gave instructions on how to avoid detection. It was, he said, his thanks to the site for enabling him to manage his pain and addiction, since he could not obtain his medicines any other way. It was either a fantastically subversive act or a cunning black-ops move. It read like both, just to complicate the matter.

The Silk Road is the most popular of the growing hidden network of drug dealers who use Tor, or The Onion Router network, an alternative web-like space that swarms with users in virtual tunnels beneath the everyday web. One drug-dealing site found there, the General Store, is even more bare-bones than the Silk Road, but still delivers the goods it offers: ketamine, DMT, MDMA. Another site, Black Market Reloaded, operates on similar principles and offers the same drugs and services as the Silk Road, though it's far less busy. The value of the service provided by the Silk Road is proportional to the numbers of people using it, and the site is quickly growing.

How does such a service as the Silk Road even continue to exist, when it is breaking the law in such a flagrant manner? In order for its customers to be completely untraceable, and therefore invulnerable to legal prosecution, the Silk Road is hosted on a hidden service, buried away on the Dark Web, far from the reach of Google. Sites on this network have randomly generated addresses, made up of a string of meaningless characters, and ending in .onion rather than .com. Its owner and its users – both the dealers and the customers – have complete anonymity. The location of the server that hosts the site is unknown, and unknowable. And, extraordinarily, the American Navy is, in some small and unintended way, partially responsible for this state of affairs.

The Silk Road is hosted on the Tor network, which allows users to browse access sites known as 'hidden services' anonymously, via a layered network of volunteer servers that encrypt traffic. As the Torproject.org website describes it, 'Tor is free software and an open network that helps you defend yourself against a form of network surveillance that threatens personal freedom and privacy, confidential business activities and relationships, and state security known as traffic analysis.'

Information activist Andrew Lewman lives between the US and Iceland, and is the mouthpiece of the Tor organization. He laughs as he recalls conversations he has had at conferences in the last eighteen months since the Silk Road began to become populated. 'People have come up to me and said: "Wow, thanks for the Silk Road!" I've been like, "Woah! Assume we're being recorded here! We don't host Silk Road – it's just an address. [Tor hidden services] are just an algorithm that provides an address."

The Tor software and network was created in 2001 by two

computer science graduates at the Massachusetts Institute of Technology. They took a piece of undeployed software, which had been written by the American Navy in 1995 to enable simple, anonymous internet use, and released their own version of it online, with the Navy's permission. 'The Navy had this project called Onion Routing, and it's still going today,' explains Lewman. 'Its goal is to defeat network traffic analysis, which is the ability to know who you are, who you're talking to, and how much data you send and receive. If you think of envelope data from your postal system, that's the basis of intelligence gathering. For whatever reason, the Navy wanted this technology – they started the project [in-house] but they didn't have any intention of releasing it publicly,' Lewman explains. 'So Paul Syverson, a mathematician who's still the core researcher for onion routing for the Navy, met grad student Roger Dingledine at a conference. Roger said, "Have you ever thought of putting this on the internet?" At the time the Navy had no plans for deployment. But Paul said sure. So the first problem to solve was that if the Navy were to release this code online – to release it to the world – they had to give up their anonymity,' said Lewman. The impossibility of having an anonymous network owned by the Navy was a comical catch-22. 'You can't have a Navy anonymity network, because no matter what individual soldier you are, your adversary will still know you are the navy and will still treat you as such,' he said.

If an anonymizing service is used by just one specific category of people then it's easy for observers to tell who they are. For example, if no one had white cars except the police, every time you saw a white car, you'd know it was the police. You wouldn't know which individual policeman it was, but you'd know they were a policeman, and would be able to see where they were going, and how often. The more people who used the system the better – the crowd offers greater cover.

The original aim of the grad students, Roger Dingledine and

Nick Mathewson, was to give users control over their data when they went online. This was during the first dotcom boom, and many companies were giving away services for free – or rather, in exchange for your data and your browsing habits, which they would then sell on to third parties. Information activists rejected that business model and wanted to offer an alternative, and so Dingeldine and Mathewson created a variant on the Navy protocol, calling it Tor. The way Tor works is best described in simple terms. When you type a web address into a standard browser, such as Firefox or Internet Explorer, your connection to the internet originates from and returns to a unique address, known as your Internet Protocol, or IP, address. This information is included inside the packet of data that you send when you press enter. The request, or packet, is then sent via the quickest possible route to the address you have specified, and then the request is delivered in the same way, but in reverse. The request for the information, and the data you receive, is stored by your Internet Service Provider, or ISP, and can be observed at many points along the entire transaction. Your ISP assigns you an IP address, which is a string of digits separated by decimal points. The address may be temporary; in some places it changes every day, in others it is active until you lose power for more than eight hours. Your IP address is the way your information requests and responses route through the network, and it can be thought of as your digital home address.

When a British net user does a Google search, the ISP routes it to a Google server, for the sake of illustration, in the US. The traffic goes across the ocean to Google and it sends all your data back to your IP address. Google now knows where you are, because of where those addresses are assigned. There are geographic IP databases that will map your IP down to a street in many cases, or at least a neighbourhood – something not desired by people who want to buy and sell pounds of hashish online. Your ISP gets to see all of your traffic and everything you do across it, along with everyone else between you and your destination, which in this case is Google, which gets to see where you live in the world so it can target ads at you.

'My IP address at home maps me down to my street,' says Lewman. 'They know exactly where I am and where I live, what street and they can probably guess what house would be mine.' Once you download and install Tor, he explains, a browser window like any other opens, and you type in addresses. It then creates a virtual tunnel from point to point, and hides each piece of data inside a series of encrypted layers, like the rings of an onion.

'What Tor does is build a tunnel which connects through three different relays in the world, so though you may physically be in the UK, your first connection may be to Hong Kong, your second may be to Argentina, your third will be in Japan. If you think of driving a car into a tunnel – in Tor you enter the tunnel in the UK, and then pop out in Japan. In order to watch what you're doing on the net authorities would have to watch the entire internet,' says Lewman.

The network and the software, which is distributed free at Torproject.org, grew with the help of funding from Voice of America, a trusted global American news site. The service approached the Tor project in 2006 and said they had noticed that users from all over the world in repressive regimes were using the Tor software to connect to their web pages. They asked the Tor volunteers to form a company, in order to make the service and the network more widely available, which they did. 'At the time we were still independent contractors with the EFF [the online freedom of speech group, the Electronic Freedom Foundation, set up by .alt newsgroup creator John Gilmore] and the American Department of Defence at the same time, which made for some strange meetings!' says Lewman.

Tor is not just used by those engaged in illicit activity. The vast majority of Tor users are simply people who want privacy when they go online, as the information gathered on us by search engines and social media grows daily. When researching sensitive or medical matters, some users don't want Facebook or Google searches sending unsettlingly accurate adverts back at them. There were thirty-six million downloads of the software last year, though that does not necessarily translate to daily users, of which there are around one million. And in repressive regimes such as Iran, Tor users can access sites that are blocked by the government. 'In Iran, between 60,000 and 100,000 people use Tor daily, for the most part [for] looking up innocuous stuff such as celebrity gossip – [but it is also used by] political activists looking to get their messages out of the tightly controlled netspace of the repressive regime. Iran's citizenry have a long history of circumventing censorship, from back in the 1960s when shortwave was banned and they figured out a way to get around it. Then it was satellite TV, but you had to get the state broadcast; Iran jams other signals. People have got good at hiding their illegal dishes, or have made them portable so they can watch TV and take it down quickly,' says Lewman. He continued, 'Now the net is just another way. And as more people go online, and get their news online, and their life is online, the government of Iran is trying to block that to maintain their censorship regime, but the Iranians are well trained in circumventing that.' The Tor software is smuggled into the country and distributed samizdat among users, by the so-called 'sneakernet' of friends walking between houses with the code on USB flash drives, disguised in encrypted files on camera cards, or buried between the etched grooves of an officially allowed CD-Rom.

Of course there is a sinister side to this libertarian technology. Dig a little deeper around this hidden, or Dark Web and you'll find pages that would give even the most extreme libertarian pause for thought. The Quick Kill page offers to 'remove the problem in your life' – for a payment of US$10,000 up front and US$10,000 once the target is eliminated. 'We are here to do business' say the site's owners, reassuring – or disappointing – prospective customers that they will not kill political figures. It's probably a scam, but it's a disturbing one nonetheless. Hidden deeper in the layers of these .onion sites are weapons dealers who tell customers looking to spend less than US$10,000 to look elsewhere. They draw the line at biological weapons.

'Somebody showed me a forum run by the Italian mafia on Tor and they traffic weapons and drugs and tonnes of garbage and toxic waste on a BBS,' says one anonymous interviewee. 'It's hidden from public net, but it's out in the open. They don't use any code words, and they have the same juvenile jokes you'd see on a usual bulletin board system.'

Lewman is realistic about the fact that the network can be used by criminals, child pornographers, drug dealers and fraudsters. 'Tor is just a technology. Silk Road and these other things would exist on something else if it wasn't for Tor. I don't know why they picked Tor and I don't care. Our code is all open source, everything we do is open source, and is mirrored all over the world. So even if for whatever reason, let's say the paedophile-terrorist-druglords and the four horsemen of the apocalypse take over Tor and that's the majority usage, then the current Tor network could shut down, and just like a phoenix it will get born again. Then maybe we'll have 10 or 1,200 Tor networks because everyone starts running their own.'

Drugs, you might consider, are the least of the authorities' worries when it comes to the hidden underbelly of the net. But not so: in June 2011 the US demonstrated that it is not content only to fight endless and expensive wars in real life, but that it now intends to take its pyrrhic battles online. That month, Democratic Senators Charles Schumer of New York and Joe Manchin of West Virginia wrote to Attorney General Eric Holder and DEA Administration head Michele Leonhart calling for the Silk Road to be shut down. Their startling lack of insight into how this part of the internet actually works might be forgivable in an uninterested or general web user. For legislators lobbying for its closure to fail to understand that the site is not findable, and even if it were found, could simply resurface elsewhere on the Dark Web, virtually guarantees that the network, and others like it, will exist, grow and gain strength for many years to come – for as long, in fact, as drugs are illegal. There really isn't any way to shut down the Silk Road unless multiple governments synchronize a worldwide jam of the entire internet – as Egypt did for a few brief days in January 2011 during the Arab Spring revolution. It soon came back online when businesses started losing money.

'All the war on drugs does is knock off the idiots on the corner because they sell it to undercover cops,' Lewman says. 'The big drug cartels can afford submarines and planes, and bribe entire police departments, which means the money is flowing somewhere. The DEA are going after the humans.' He says the DEA's interest was piqued by Senators Schumer and Manchin's bluster. 'In a long line of things that will kill America, Silk Road is the worst right now. That got a lot of attention and press but had the opposite effect to what they wanted. What people heard was that Silk Road has really good drugs!' he says.

Silk Road's owners are entirely anonymous. But the name of the current site administrator is an intriguing clue as to the way the service may be run. Though Dread Pirate Roberts was previously named Silk Road in the market's attached forum pages, and Admin in the marketplace itself, he renamed himself in February 2012, delighting in posts for days beforehand in the appropriateness of the name he would soon reveal. It turned out he had chosen to call himself after a swashbuckling pirate character in the 1973 fantasy novel *The Princess Bride* by renowned American screenwriter William Goldman. In that work, the Dread Pirate Roberts is a persona assumed by many different characters, each of whom hands the mantle, name, responsibilities and ship to his chosen successor. It seems possible that the Silk Road site is run not by a single operator, but by a loosely tied conglomerate of individuals, each of whom successively takes the risk – and reaps the rewards – that running the site must entail. It's impossible to say. Perhaps the name more simply nods at the popular respect for the outlaw-pirate hero figure who has thrived in a world where torrent downloads are far more respectable than shoplifting a DVD.

Appropriately enough, it was William Goldman who wrote the line 'Follow the money', in the 1976 film *All the President's Men*, a phrase that has become the battle-cry for investigative journalists looking for examples of official corruption.[2] The architecture of bitcoin, the currency used on the Silk Road by dealers and users, and other services deployed by the site, mean the money *cannot* be simply followed. Transactions are almost anonymous, and communications are encrypted between the intended recipients, making eavesdropping impossible.

Proof that the site delivered was broadcast in the UK in February 2012 when researchers from BBC Radio Five Live ordered and received a sample of very high-purity DMT. The sample was tested by John Ramsey at St George's, and he said it was indeed excellent quality. The story was broadcast and published on the BBC's website. Andrew Lewman audibly facepalms as he relates the story over the telephone. 'What better advert could they have given? Not only does this illegal site sell rare drugs, it sells very high-quality product.' But you didn't need to trust the BBC. The forums at the site offered crowdsourced proof of the best vendors and worst scammers. In June 2012, reviews for the best LSD vendor ran to eighty-one pages, with 50,000 views, heroin to twenty-two pages with 8,000 views. Cocaine vendors were reviewed in a 292-page behemoth with over 90,000 views, and MDMA ran in at 129 pages with over 60,000 views.

One vendor said dealing drugs on the site wasn't without its moral problems. 'The prospect of a twelve-year-old loaded to the gills on my MDMA is not a pleasant one. Enabling self-destructive/addictive behaviour is also upsetting to me. Dealing in real life [IRL] you can recognize abuse and let customers know you're concerned, but online, there's no way to tell.'

He admitted, though, that vending on the site was financially much more lucrative than selling in real life: 'IRL, you're limited by your social circles, but here it's only a question of supply, capital and hours in the day. Packaging straight-up sucks to do. It's extremely monotonous and requires a good degree of concentration to avoid making any mistakes that might endanger the customer receiving. Sometimes during especially busy periods I spend 70, 80, 90 hours a week packaging, all of it extremely dull. Apart from the risk of being locked up for the next decade it's definitely the worst part. Dealing in real life is much more pleasant.'

Greater paranoia about the authorities is another downside: 'Public drug markets [such as this] are a giant middle finger to many powerful interests and so the political motivation to shut them down and lock up the people participating is out of proportion to the actual volume of illicit trade taking place. Last summer I was the "number one" (basically highest-volume) vendor on the site for a while, and the fear really crept up on me. I'd lie awake at night thinking about it, worrying I was going to have my door kicked down and be dragged away at any moment. I'm much more comfortable with it now, but if I had known from the start how much mental torment and stress were involved with vending, I probably wouldn't have started.'

However, there are upsides, he says: 'I find the day-to-day grind of vending online worse than dealing IRL, but the human interaction online is often a lot more uplifting in some ways. Most people I sell to IRL are club kids/raver types so they're more predisposed towards hedonism (which I of course have nothing against!) than using for more spiritual/emotional reasons, so the feedback is less touching, which is a definite negative for me. I get emails from Silk Road customers telling me how the drugs I sell have helped them with emotional or spiritual or sexual problems, people mending broken relationships, rekindling intimacy.'

If predictions that sites such as the Silk Road will become more popular, or even commonplace seem far-fetched, think back just seventeen years. At that point, Amazon.com was a three-person startup launching from a garage; today, it's the first place most people look to buy almost any object that can be delivered.

And exactly the same issues confront users of online drugs markets today as faced those who dared enter their credit card details on to a bookshop's website in the 1990s. Will the vendor deliver? Can I trust this software with my money? The only difference is that Silk Road customers might find themselves wondering whether their purchase will result in a jail sentence.

The motivation for people to use the Silk Road is high, given the prevailing legal climate. Considering that the Royal Mail in the UK delivers 15.9 billion items a year to the UK's 29 million addresses[3] and that small envelopes and packages are seldom opened, much less X-rayed or sniffed by dogs, capture, prosecution and imprisonment look unlikely.

One vendor on the site even offers a fake package service for the super-cautious: he'll deliver you an empty box or envelope for a small charge, just to get the postman used to delivering packages from overseas.

Packaging by many vendors on the site is said to be exceptionally ingenious, and the protocol on the forums and in feedback forms below purchases is never to discuss the details of these. What's more, there are vendors in many countries so there's no need to worry about international postal or customs issues: users in the US or UK or the Netherlands – or indeed, in dozens of countries worldwide – can buy drugs from dealers in their own countries, removing the danger of border staff targeting their packages.

Technically, while the Tor network is now small, meaning its pages load more slowly than those on normal websites, in coming years the power of cloud computing means that more relays carrying the service will be able to be set up cheaply. In November 2011 Amazon's cloud servers started hosting Tor bridges. For three dollars a month, users click and support the project, with no knowledge of the technology required. And in late 2012, the Noisebridge group of online activists made supporting the Tor network as easy as clicking on a donate button at Noisetor.net. Anonymity 2.0 – click here to buy now.

Most politicians speak as if they believe there is an 'off' button for the net that can be thrown without affecting business interests, too. But the dreams of early net pioneers, for better and for worse, are now coming true.

People are now connected to each other with no central hierarchy governing that process; information flows freely and respects no authority, and the network is indestructable.

As Roy Amara, a futurologist and Stanford computer engineer, and president of Institute for the Future, said: 'We tend to overestimate the effect of a technology in the short run and underestimate the effect in the long run.'

A new kind of currency is making official control of this area even harder. Bitcoin is an electronic cash system, produced using cryptography. It is a peer-to-peer currency, made by users, meaning that no central authority issues money or tracks transactions. For every legal bitcoin user, selling web design services or carrying out coding jobs for which they are paid in the currency, there are many more using bitcoins to buy drugs on the Silk Road. Bitcoin is today the preferred choice of hundreds of online drug dealers. You can buy bitcoins using cash or other currencies in hundreds of ways, with varying levels of anonymity. Using bitcoins can be, depending on how you use them, almost completely anonymous.

Originally, bitcoins were produced by 'miners' – a figurative term for computer owners who donated their processor time to the project and were rewarded with coins for their efforts. The currency, or rather, the system that creates the currency, was released to the web on 1 November 2008, as the world economic system teetered on the brink of systemic collapse. Anonymous software coder Satoshi Nakamoto issued the open source application, and included a sly reference to the latest banking bailout by Britain's then-chancellor of the exchequer, Alistair Darling, buried in code for the so-called Genesis Block – the first coins ever 'mined', in January 2009.[4]

The main reason no purely digital currency has ever gained traction is because data-as-cash has a central flaw. As the music industry has discovered in recent years, digital information is infinitely copyable. Digital money, until now, could be spent again and again. To prevent this, a central banker would be required, someone who would maintain a sales ledger. But who could be trusted with such a thing? Bitcoin solved that problem by turning to the crowd for the answer, distributing the record of transactions much like a .torrent file.

Torrents are shared downloads, so when a user fires up their bit-torrent client and downloads, say, a film using a torrent file from an illegal site such as the Pirate Bay, they actually download millions of chunks of the file from a swarm of users at once, rather than one file from one central server. The torrent software on the downloaders' machines then assembles the pieces of data into a film or music file.

Nakamoto's elegant solution to the double-spend dilemma was to create what he called a 'block chain', a distributed, or shared ledger of all transfers of coins from one person to another. Crowdsourced, decentralized, massively distributed cryptographic cash had arrived.

Users, known as miners, donate processor time to maintain and update the block chain, which records all transactions between users, and in the process also 'dig' for new coins. Miners' computers send evidence of those transactions to the network, racing each other to solve these irreversible crypotographic puzzles that contain several transactions. The first miner to crack these puzzles gets fifty new bitcoins as a reward, and those transactions are added to the blockchain. The puzzles are designed to become more complex over time as more miners come on board, which maintains production to one block every ten minutes, keeping the creation of new coins steady. The reward for successful mining also falls over time, from fifty to twenty-five coins per block, and drops sequentially by half every 210,000 blocks. In the year 2140, there will be no more bit-coins minted or mined − the software limits their production, meaning there will only ever be twenty-one million coins in existence, preventing inflation. They can, though, be divided to eight decimal places, with each sub-unit known as a satoshi, after the coder who invented them.

Bitcoin could almost be seen as performance art; it demon-strates in the most practical way what many people have never considered: that the system of money, of currency issuance, is illusory at best, deceptive at worst. As if to illustrate this, one truly psychedelic item was put up for sale on the Silk Road in July 2011, when a vendor named Uglysurfer offered pound weights of American copper pennies for ฿10.43/lb. The face value of the pennies was US$14.60, but at the time, copper prices were such that the metal contained within a American penny was worth almost three cents. Uglysurfer was demonstrating that the 'fiat' system of money and fractional reserve banking, whereby banks can and do create money from thin air, was not to be trusted.

'Under the best conditions, I could walk into a bank and provide US$25 dollars in paper Federal Reserve Notes, and walk out with a box of 95 per cent copper pennies with a metal value of approximately US$64 (in copper),' he explained. 'Not a bad deal! Of course not all of the pennies will be 95 per cent copper, but the portion of 95 per cent copper pennies in the box have the proportional gain,' he told me. 'I have to sort the copper pennies from the zinc pennies and I have designed a system to automate the sorting based on pattern recognition of metal composition. So I use technology to sort and reach a scale of efficiency that makes the process profitable. In my opinion, fiat currencies are doomed simply because of the deception involved. As the populace is educated (and it looks like that education is about to be painfully forced on the masses – look at Greece) it will be a force of nature – the destruction of the fiat model. Anonymity or the ability to act anonymously is a critical means to preserving individual freedom in the midst of tyrants. I am a true believer in financial privacy. My belief is today those that seek personal freedom become enemies of the state (as far as the state is concerned) and are in the eyes of the state criminals. Not unlike those who deal in drugs on Silk Road.'

In 2009 Laszlo Hancyez, an American programmer, made the world's first purchase using bitcoins.[5] He sent the bitcoins to a British man who called in a credit card payment transatlantically. It was a pizza, and it cost ฿10,000 – a sum worth £75,000 in November 2012. Today, many thousands of bitcoins are circulating around Silk Road users, and around 12,000 per day are spent on the site, at a value in late 2012 of around £7.50 each.

Silk Road users value the currency for its supposed anonymity, although it is not entirely untrackable to the curious and competent, nor is it entirely safe. Live by the chip, die by the chip: in June 2011 a user named Allinvain claimed that 25,000 coins had been stolen from his computer. A week later, a hacker compromised security at MtGox, a Japanese firm which handles the vast majority of cash-to-coin exchange, and pretended to be selling off a vast chunk of the currency. As a consequence the price dropped to zero, allowing him to steal thousands of coins. The system was then flooded with speculators, forcing MtGox to limit withdrawals to US$1,000 worth of bitcoins a day to stem the flow and prop up the dollar-value of the currency.[6]

Network analysts Fergal Reid and Martin Harrigan of University College Dublin wrote a 2012 paper baldly titled 'Bitcoin is Not Anonymous'. In it they demonstrated what the high-tech coining community knew – that the blockchain recorded all transactions. Reid posted in a comment thread following the release of his paper, 'You don't get anonymity automatically from the system. A lot of people out there think you do.'[7]

But the determined user can retain anonymity easily enough in the US at least, by entering a bank and paying cash into an exchanger's account, for bitcoins are now traded just as dollars and euros are. (They now have a value that is decided by the market. The total bitcoin market capitalization stood at £72 million in November 2012 – with around 10 million coins valued by the secondary market at around £7.50 each.) By this method, cash exits the real world, and from there can enter the miasmic smog of this market.

Bitcoin addresses are generated anonymously and instantly, and infinitely. You can launder bitcoins bought with pounds from your bank account and send it through 100, or 1,000, anonymous bitcoin accounts that you have generated and which you control in just a few hours, then use them to buy drugs. There is no trace, especially if you connect to the net with Tor.

And there are many services online where users can buy other digital currencies, and convert them into bitcoins. Liberty Gold is a virtual metal-backed currency from Costa Rica, purchasable automatically from anonymous servers with Western Union cash payments, whereby participants swap the transaction number for invisible currencies which they can then swap into other currencies. You could for a short period in 2011 even buy bitcoin by SMS: users would buy a simcard from Poland, or Belgium, or one of a dozen other countries, charge it with cash, send a text and receive their coins to their handset. 'Mixing' services too, can tumble the coins in and out of thousands of other bitcoin transactions and accounts, making a dense web of mathematics even denser still. When most investigators can't even understand the basics of encryption, the likelihood that they or a jury member will reach an understanding of bitcoin is minimal.

And when most small-scale drug transactions are small, under £100, who's watching? The answer, so far, is that no one has been busted using evidence from the bitcoin blockchain. Bitcoin addresses, where you receive and store coins, are randomly generated strings of letters and numbers, and there's no ID check system – and you can create another in moments. If that's not enough, the more paranoid users can use a service such as Bitcoinfog, which matches deposits and transactions randomly, paying out the total you paid in in a series of different amounts. Then there are instawallets, temporary, one-time-use holding accounts where coins can be stored for a few seconds over an anonymized net connection and spat out elsewhere. Or there's Coinapult, a jokey service allowing users to sling coins to each other across the ether. There are games such as Satoshi Dice, a gambling game that allows micro-bets on random chance algorithms. Since the currency is divisible to eight decimal places, the thousands of tiny bets further complicate the block chain and disguise criminality.

There's no denying that this is a minority sport, and that the process is arduous, and can sometimes fail completely. Online wallet services, where coins can be stored on the net, rather than on your computer's hard drive, are often scams that can easily fleece users. The complexity of the system does not lend itself to the kind of impulse purchase made by some drug users. But that hasn't stopped thousands of users of the Silk Road from embracing the technology. Networks grow and proliferate if they are populated, required and scalable. Bitcoin, Tor and the Silk Road fulfil all of these criteria.

Might this arcane and hidden world spawn new and different versions of itself? Those who believe this system is so complicated that it will never catch on might perhaps consider that within living memory, even configuring basic internet access took expert knowledge. Nowadays, we only actually notice our net connections exist when they drop.

Encryption is what makes this market possible, and what makes it so hard for lawmakers to attack. Encryption works by scrambling information and only allowing the holders of two sets of keys to decode that information. The public key is known to everybody and is published. The secret key is held only by the recipient. Alice wants to tell Bob some sensitive information – or indeed any information intended only for his eyes. So Alice uses Bob's public key to encrypt the message to him. Bob uses her private key to unlock, or decrypt the information. No one else can read it.

In a 1991 paper, Phil Zimmermann, coder and security special-ist, and author of the software package Pretty Good Privacy, wrote:

> It's personal. It's private. And it's no one's business but yours. You may be planning a political campaign, discus-sing your taxes, or having a secret romance. Or you may be communicating with a political dissident in a repres-sive country. Whatever it is, you don't want your private electronic mail (email) or confidential documents read by anyone else. There's nothing wrong with asserting your privacy. Privacy is as apple-pie as the Constitution. The only way to hold the line on privacy in the information age is strong cryptography.[8]

If governments or police wanted to read the messages between Silk Road users, they'd have to spend years in so-called 'brute force' attacks, where hundreds of millions of possible passwords are tried one after the other.

In the UK, though, if you are investigated by police and use encryption, and refuse to give your passwords to investigators, you will be charged with a crime and jailed under the Regulation of Investigatory Powers Act (RIPA). No matter what your defence, no matter what crime you are under investigation for, even in the absence of any other evidence, if you maintain your right to private communications, you will be deemed a criminal and jailed.

IT website *The Register* reported in 2009 that the first person jailed under part III of the RIPA was 'a schizophrenic science hobbyist with no criminal record'. Found with a model rocket as he returned to London from Paris, he refused to give police the keys to his encrypted data, indeed, he refused to speak at all, and was jailed for thirteen months. Six months into his sentence the man, named only as JLF, was sectioned under the Mental Health Act and does not now know when he will be released.[9]

It's highly likely that legislators will one day use the menace of online drug deals as a justification for intruding into people's privacy. A happy consequence for the government of its targeting of this straw man folk devil will be unfettered access to all our private thoughts and conversations.

You can never be sure a conversation is private without encryption, John Callas, an American computer security expert who co-founded PGP Corp with Zimmermann, tells me. The German government broke Skype's encryption models by releasing malware and viruses into the wild that can easily unscramble voice calls across the network, allowing it to eavesdrop at will, he tells me – across a Skype line. 'In the old days, hundreds of years ago people could speak privately by going out and taking a walk around the green and talking among themselves and there was no way people could listen in,' he told me. 'Today [with long-distance communication so commonplace] there's no good way to do that except by using technology. Encryption lets you have a private conversation with anyone else, and that's needed by business and anyone that wants to talk in private.'

The history of encryption is a fascinating tale of early net privacy campaigners facing down the government – and winning. From the 1970s onwards, encryption was considered military hardware and could not be exported from the US. In 1995 Phil Zimmermann had the source code for PGP printed in book form and sent to Germany from the US, since the export of literature was not banned. An engineer in Germany scanned the code, recompiled it and distributed it online. The export regimes were eventually liberalized, as the government had to accept that encryption was nothing more than maths. 'These networks were not designed to respect orders,' deadpans Callas.

Could governments roll back encryption advances in order to prevent online drug dealing, and halt secret communications? 'I think the toothpaste is out of the tube,' says Callas. 'Cryptography, in some form, is used by people every day all the time. Whenever you buy something online, your purchase details and delivery details are all encrypted. There are reasons for that – there are gangs that want to steal your info and defraud people with it. The reality is that among the other problems society has, including the Mafia stealing from old ladies, the way to protect them is encryption. It is flat technologically impossible to manage encryption,' he told me.

Callas is certain government will focus on the drugs issue in the upcoming debates around encryption and privacy. 'Encryption is why the big NSA [National Security Agency] facility in Utah is being built. The NSA understands it is a new century and they need new technology for what they are doing,' he said. The new NSA facility is a data-harvesting plant in the desert near Utah. It will cost two billion dollars to build, will measure a million square feet, and will be able to store 500 quintillion pages of information. It is Callas' belief that this centre is being built for traffic analysis purposes – seeing who is talking to whom, how often and for how long – and to engineer password-breaking technologies. Though encryption is essentially uncrackable, passwords are generally trivial to break. Traffic analysis can also be used to gather valuable data on communications that have passed through Tor.

In just under two years, the Silk Road administrators have used technology and ingenuity, along with innovative crowd-sourcing solutions to internal and external threats, to achieve what thousands of campaigners have toiled since the 1960s to achieve: the right for people to buy and sell natural and artificial chemicals that affect their consciousness in ways they choose without interference from the state. It is a paradigm shift that cannot easily be reversed.

The growth of Silk Road may have provoked the very public forced closure in 2012 of one of the net's longest-standing online drugs markets. The Farmer's Market, or TFM as it was known, was an accident, or more accurately, a bust, waiting to happen. The site operated for a number of years as an email-only service at adamflowers@hushmail.com. Later, it ran its business on the anonymizing Tor network but, foolishly, even there, used the Hushmail encrypted email service to serve its thousands of international customers rather than using its own encryption. The site sold mainstream psychedelics – MDMA, LSD, ketamine and high-potency marijuana and hashish, along with DMT, psilocybin mushrooms and mescaline. Its vendors were connoisseurs, and offered rare cannabis strains and seeds seldom available anywhere else. It was mainly a boutique online marijuana store and its descriptions showed the expertise of the obsessive.

TFM's existence was an open secret in the online drug dealing and purchasing community – far too open. The only startling thing about the closure was that it took so long. The site was like a proto-Silk Road, but crucially, as court papers would reveal in 2012, it accepted payment methods that were traceable and insecure. Users could join the site with no invitation, and therefore with no background or reputation checks. With its drop-down menus and creaky lo-fi design and jagged fonts, it felt a rather rustic kind of place, an artisanal street market – if street markets had rickety oak barrels filled with pounds of free-flowing crystal ketamine and fragrant sprigs of marijuana rather than single-estate coffee beans and overpriced sourdough bread.

The bust came one year after mainstream media outlets became aware of the Silk Road's existence; it was a showboating exercise to satisfy political pressure from the US to do something about the new internet drug menace – and maybe to scare off users from buying drugs online generally. The bust was hardly hi-tech, nor was it particularly ingenious, however much the police attempted to portray it that way. Undercover agents infiltrated the network posing as buyers, and simply made orders that revealed the network's international links, names, bank account details of the recipients of funds, and addresses connected to the dealers.

Every criminal enterprise has a weak point, and one of TFM's most fundamental errors was that it took payment via various insecure and far-from anonymous means, from PayPal to Western Union international transfers. They also accepted I-Golder, a digital gold currency, and Pecunix, a similar currency, which stores its ingots in Swiss vaults but is incorporated in Panama, the Central American banking powerhouse that borders Colombia to the south and east.

The Pecunix payments were laundered through various PayPal accounts, and then sent through various accounts in Hungary, Western Union payments skipping across continents to become balances on-screen in I-Golder and Pecunix accounts, and back, and forth until TFM's dealers thought, mistakenly, that the money was crisply laundered. They were wrong. The transactions had been tracked through these systems; the paper trail was easy to follow. If they'd used bitcoin, the site's operators would be free men today.

The indictment alleged that between January 2007 and

October 2009, The Farmer's Market processed 5,256 orders with a value of US$1.04 million. The site had over 3,000 customers in thirty-five countries, including buyers in every state of the USA. Forty-two-year-old Marc Willem, the lead defendant also known as adamflowers, was arrested on 16 April 2012 in Lelystad, Netherlands. The day before, Michael Evron, an American citizen living in Buenos Aires, Argentina, was arrested as he attempted to leave Colombia. Six other dealers and accomplices were arrested at their homes throughout America. At the time of research, none of their 3,000 customers had been targeted. The indictment ran to sixty-six pages, and documented hundreds of drug deals that the group had administered. The network was huge, covering countries in Central, Latin and North America, Eastern and mainland Europe. The men were charged before the United States District Court for the Central District of California on charges of conspiracy to distribute controlled substances, conspiracy to launder money, distribution of LSD, aiding and abetting, continuing criminal enterprise and criminal forfeiture. When the news broke online, panicked chatter spread across dozens of sites.

Police called the group 'sophisticated' and said it used 'advanced anonymizing online technology'. This was not true; the group used proprietary encryption on a webmail service, Hushmail, which publicly stated it would cooperate with police if asked to do so. The 'advanced anonymizing software' was simply Tor, which, though it is indeed advanced, is something even the most technically illiterate web user can use easily. Briane Grey of the DEA said the operation – named Adam Bomb – 'should send a clear message to organizations that are using technology to conduct criminal activity that the DEA and our law enforcement partners will track them down and bring them to justice'.[10] The police's intention was to give the impression they had in-filtrated an elaborate and complex market; they had not – they had just sent a few emails and issued a few subpoenas to follow the money.

In the final analysis, TFM was low-hanging fruit for the police, and in grabbing it, they merely showed their hand early and revealed the weak points of any online peer-to-peer drug smuggling network – communications and payment. The slew of news stories also told anyone who was listening that it was possible to buy drugs online, and that it was the dealers, rather than the site's 3,000 happy users that the police were targeting.

The Silk Road's payment and communication systems remain essentially impenetrable. It's here on the Silk Road that the early net evangelists' vision of a world where information flows freely, where no central hierarchy rules, and where the network takes precedence over the individual has finally been realized. Whether you celebrate or lament the fact that drugs such as cocaine, heroin and LSD are now available online with just a little effort and very little likelihood of legal consequences, it is undeniable that we are at a turning point in legal history.

Through a decades-long process of chemical and technical innovation, drug users and producers have beaten the laws made by a political system whose only response to increased drug use is a harmful, expensive, counterproductive and ultimately failed strategy of criminalization.

Over the course of the century or so that drug laws have existed in any meaningful form, a clear pattern has emerged. As each law to prevent drug consumption is made, a means to circumvent it is sought, and found. Those means can be chemical, legal, social or technological. We stand today at a crossroads formed by those four elements, with the web making possible communication between distant strangers, facilitating the sharing of limitless quantities of information, and enabling the distribution of drugs anywhere in the world. Where do we go next?

# 10

## Prohibition in the Digital Age

In 1920, the American government banned alcohol sales, urged on by the Church. During the first two years of Prohibition, consumption dropped at first, but then increased enormously – insurance companies said the increase in alcoholism tripled. 'Speakeasies' or illegal bars flourished and in New York alone, there were soon 30,000.[1]

Efforts to stop the smuggling of alcohol from neighbouring countries stepped up, but organized crime syndicates took control of the newly lucrative trade and soon gained the working capital required to finance other schemes, such as casinos. As the borders shut, criminals sought alternatives and stole vast quantities of industrial alcohol used in the chemical industry, redistilling it to rid it of contaminants and make it fit for human consumption. The government, angered that its laws were being flouted, decided to poison the alcohol to scare people out of drinking it. In a largely unreported act of deliberate mass murder, the government added new, more dangerous substances to it. A *Slate* article in 2010 listed these as 'kerosene and brucine (a plant alkaloid closely related to strychnine), gasoline, benzene, cadmium, iodine, zinc, mercury salts, nicotine, ether, formaldehyde, chloroform, camphor, carbolic acid, quinine, and acetone, as well as methyl alcohol.'[2]

In December 1926, twenty-six people died in a matter of

days as they celebrated Christmas with the poisoned booze. The government was not legally responsible since it had banned alcohol sales and the use of alcohol, however it was obtained. In its view, people were choosing to poison themselves. 'The government knows it is not stopping drinking by putting poison in alcohol,' New York City medical examiner Charles Norris said at a hastily organized press conference. 'Yet it continues its poisoning processes, heedless of the fact that people determined to drink are daily absorbing that poison. Knowing this to be true, the United States government must be charged with the moral responsibility for the deaths that poisoned liquor causes, although it cannot be held legally responsible.'

Who then is morally responsible for the sickening of the thirty young Alabamians who are today suffering kidney damage after smoking tainted Spice products? Legal responsibility lies with the manufacturers and retailers of the fake pot products that were tainted with herbicide as well as the research chemical that made their customers ill. Personal responsibility lies with each of the young men and women who chose to smoke the tainted herbal blends. But in a country where 850,000 people have been arrested for cannabis use since erstwhile marijuana enthusiast President Obama came to power, where does moral responsibility lie?

It is mainly the young who are suffering the consequences of society's inability to update our drug laws effectively for the modern age. Almost one third of young people are searching for ways of getting legally high, according to the latest survey commissioned by the Angelus Foundation, a campaign group founded in 2009 by Maryon Stewart, whose twenty-one-year-old daughter Hester, a gifted medical student and keen athlete, died after taking GBL in 2009. (Gamma-butyrolactone, a paint stripper and industrial cleaner, can be used as an intoxicant and is poplar on the club scene. It is active at 1 ml, and causes euphoria and disinhibition, but overdoses, where users fall into a coma-like state, are commonplace since it is so potent. It was legal until late 2009.)

Two-thirds of the 1,011 sixteen-to-twenty-four-year-olds surveyed by the Angelus Foundation in October 2012 admitted they were not well-informed about the risks associated with the new drugs on the market.[3]

Festivals since Woodstock have been linked with drug use, whatever message their PR machines might seed in the press, so events there can tell us much about current trends of use and the attendant problems. Dip your head under the canvas at a festival medical tent and you arrive at the intersection of the net, new drugs and young people. Monty Flinsch, who runs Shanti Camp, a non-profit aid organization providing drug crisis intervention at American festivals, says that in recent years instead of dealing with the psychological issues caused by LSD, psilocybin and MDMA, they have seen seizures, delirium, violence and deaths. 'Even discounting the hyperbolic news coverage of face-eating zombies, the real situation is substantially worse with legal research chemicals than it ever was before. It is now easier for an American teenager to obtain a powerful psychedelic than it is to obtain alcohol. Today's scene is much more complex with the influx of large numbers of research chemicals ranging from the more common bath salts (MDPV, methylone) to much more obscure chemicals such as 25C-NBOMe and methoxetamine,' he said.

The reasons the drugs are taken are manifold, but he believes their legality is a major draw, along with cultural influences. 'Kids feel they are exposing themselves to less risk by taking drugs that are not going to get them arrested, and drug use is highly subject to countercultural trends, and whatever the cool kids are taking quickly becomes popular. In many cases the legal consequences of drug use far outweigh the medical risks. Our drug laws in the US are forcing users to experiment with increasingly dangerous compounds in order to avoid having their lives ruined by a criminal conviction.'

Flinsch says he cannot see any likely improvements in the future. 'New research chemicals are ubiquitous and the problems associated with them are growing. From the frontlines we see the situation getting worse rather than better. The new compounds are poorly understood and have little or no history of human use, and therefore the problems we see are harder to characterize and therefore treat. It is sad that what is currently legal is substantially more dangerous than what is illegal.'

The entire debate around drugs, which was already philosophically and practically complex, has been made yet more intractable by the emergence of these new drugs and distribution systems. Our insistence on overlaying anachronistic models of drug control onto this digital world might, in future years, be seen as a fatal flaw that we did not address when we had the chance.

The popularization of research chemicals presents legislators, policymakers and police with an almost existential dilemma. They are charged with protecting the health of populations and reducing crimes, and these new drugs pose health risks, but are legal. The Chinese factories that produce them operate with none of the quality control typical in most pharmaceutical manufacturing plants, but customer uptake is enthusiastic. Each new ban brings a newer, possibly more dangerous drug to the market, and it is impossible to predict what the next moves might be.

Legal responses seem not only not to work, but to exacerbate the issue. The American Analog Act did nothing to prevent the arrival in 2009–11 of the JWH chemicals, the cathinones found in bath salts, and the other synthetic cannabinoids that had hit the UK and Europe in 2008. And where the early vendors of synthetic cannabis substitutes had sold the drugs online, the US did it bigger and better, and even more publicly and commercially.

In the US, in October 2011 the DEA responded by adding several of the new drugs to the controlled-substances schedule, making them formally and specifically illegal. The Synthetic Drug Control Act of 2011 was finally signed into law in July

2012, banning dozens of research chemicals at a stroke. Soon after the bill was passed, *Time* magazine quoted a Tennessee medic, Dr Sullivan Smith, who said the state had been engulfed by the new drugs. 'The problem is these drugs are changing and I'm sure they're going to find some that are a little bit different chemically so they don't fall under the law,' he said. 'Is it adequate to name five or ten or even twenty? The answer is no, they're changing too fast.'[4]

Within weeks of these laws being passed, there were dozens more new drugs available in the US. One category, known as the NBOME-series of chemicals, is composed of unscheduled analogues of the banned Shulgin psychedelics 2C-I, 2C-B, 2C-D, and so on. Where Shulgin's chemicals were generally active between 10 mg and 20 mg, these new compounds, created in legitimate medical settings for experimental purposes, are more potent by a large order of magnitude, active at around 200 µg. Each gram of these new, unresearched drugs contains around 5,000 doses, and they cost fractions of a penny per dose. The compounds existed before the most recent bans, but it was the new laws that inspired their wider use; use that will only grow as talk of their effects is amplified online. They have already claimed victims. At the Voodoo Fest in New Orleans in October 2012, twenty-one-year-old Clayton Otwell died after taking one drop of an NBOME drug. The *New Orleans Times Picayune* newspaper spoke to festival goers who said many dealers were selling the drug 25I-NBOME as artificial LSD or mescaline at the event. 'This weekend, it was everywhere,' festivalgoer Jarod Brignac, who also was with Otwell at the festival, told the paper. 'People had bottles and bottles of it; they were walking through the crowd, trying to make a dime off people at the festival.'[5]

There have been at least six other fatalities in the US from 25I-NBOME, Erowid reported in late 2012.[6] There are dozens of other NBOME-drugs, and their use is growing. The Bluelight bulletin board has three threads on 25I-NBOME, running to over seventy-five pages with more than 100,000 views. Search Google for it and there are suppliers on the first page. A kilo of it can be bought for a few thousand dollars from China.

Britain's current response to the emergence of research chemicals as legal highs in the UK is to ban each product as it appears via TCDOs, but this is unsustainable in the European context, firstly because the law is not European-wide; that is to say, the chemicals it bans are easily available from some neighbouring countries and since distribution is carried out via ecommerce and home delivery, without a nationwide domestic policy of opening or examining via X-ray or sniffer dog each envelope that arrives in the UK, the drugs will continue to enter, even at retail quantities. Let us not forget that Britain consumes a couple of tonnes a year of cocaine, and that has been banned for over a century.

Secondly, the law is unfit for purpose as it sets governments and police and forensic staff an impossible monitoring and enforcement task: at the last count, the European Monitoring Centre for Drugs and Drug Addiction (EMCDDA) found there were 690 webshops selling at least one synthetic drug each in Europe, not to mention bricks and mortar outlets on high streets.[7]

Consultant addictions psychiatrist Dr Adam Winstock says he doubts the efficacy of TCDOs longer term. 'TCDOs are a good idea in principle, but will the government invest adequately in the harm and risk assessments required? I think not. And if they did [and the new drugs the TCDOs had temporarily banned were harmless] would they then license a new psychoactive drug?'

The problem is not going away, no matter how much legislators might wish it. 'New drugs have become a global phenomenon which is developing at an unprecedented pace,' said an EMCDDA report in April 2012.[8] Director Wolfgang Götz wrote: 'We now see new drugs marketed in attractive packages on the Internet or sold in nightclubs and on street corners. Whatever the source, the simple fact is that a dangerous game of roulette is being played by those who consume an ever-growing variety of powders, pills and mixtures, without accurate knowledge of what substances they contain and the potential health risks they may pose.' He added, 'We must continue to enhance Europe's ability to detect and respond quickly and appropriately to these developments. This requires networking and the sharing of information and it requires greater investment in forensic analysis and research.'

The EMCDDA admitted that these new drugs posed health risks to individuals and the general public, and acknowledged that legal moves to ban the drugs could simply see the cycle repeat: 'Legislative procedures to bring a substance under the control of the specific drug law can take over a year in some countries. And controlling a substance may have unintended consequences, such as the emergence of a more harmful, non-controlled replacement.' Yet even while identifying the risks to thousands of drug users across Europe, and acknowledging that its strategy could lead to greater harms, it did not propose any solution to the issue other than the prohibition that the organization itself argues actually created the trade in substitute drugs.

Europol director Rob Wainwright, quoted in the EMCDDA report, said: 'The selling of illicit drugs and new psychoactive substances is yet another area where the Internet is abused by organized criminals. We must ensure that law enforcement agencies have the modern operational and legislative tools to combat such cases effectively.'

British police have different views, however. The Association of Chief Police Officers' national lead on drugs, Tim Hollis – one of Britain's most senior policemen, with a thirty-five-year career in the force – confirmed that the emergence of research chemicals and other para-legal recreational drugs is not a policing priority, and even pursuing prosecutions for possession of standard drugs isn't at times. He insisted that changing behaviours and attitudes is the best way to prevent drug use: 'Street cops recognize that if kids have drugs in their possession and you haul them before magistrates, it's a good box-ticking exercise, but does it change their behaviour? Frankly, no. The service is pragmatic about how we can change the choices young people make. I'd like to think we could have a better-informed debate on how to use scientific and health evidence to help young people make good choices, and how to help those who make choices that are inherently unhealthy for them and society.'

Hollis, who laments the inability of both sides of the drug debate to come to a consensus rather than taking up polarized and politicized positions, argues convincingly that for the majority of people who do not take drugs, decriminalization is not a viable option. 'The decriminalization and legalization approach worries a lot of the overwhelming law-abiding majority who do not use drugs. The Home Office statistics show that the reality is the vast majority don't take drugs. It's also an emotive issue: if you tell the parents of a child whose kid has died that the drug that killed them is safe, it's not going to work.'

However, this realism works both ways for Hollis, one of the UK's most senior drugs officials, who says he believes the public expect police to tackle crimes that affect communities more than targeting internet dealers. 'Are there squads of officers sitting on computers monitoring this?' he asked. 'No. The public wants visibility, police on the street – not a police member sitting in an office monitoring the internet trying to spot people [selling drugs]. This is planet earth!' he told me. 'As police, we deal with all aspects of harms, from terrorism to organized crime, to road traffic accidents, where we talk to the parents of the dead and tell them what's happened. We talk to victims of violent assault and of sexual assault in hospital. From a police perspective, how excited are we about some young people buying a white powder from a headshop, which may or may not be what it is claims to be? Is there a risk? Of course, it's a risky world. Can we [monitor] it all the time? No, it's not realistic.'

Even as the flow of chemicals continues unabated, the UK Border Agency believes jailing dealers is a deterrent. In February 2012, its investigations were instrumental in the conviction of a twenty-six-year-old Kent man, Jeremy Detheridge, for the supply of drugs sourced from China that he believed were legal to sell with the no-consumption caveat. Detheridge chose to disguise them as lawn feed. It was one of the first such cases that resulted in a conviction.

Paul Tapsell, prosecuting, said that in April 2010 two packages from Shanghai Yiyi Maoyi Co, China, were intercepted at Stansted airport. They were addressed to an alias, David Saunders in Botany Road, Broadstairs. Upon testing at the airport, the packages were found to contain lignocaine, a topical anaesthetic, and MBZP, a recently banned Class C drug of the piperazine class. Judge Adele Williams said Detheridge did not know what was in the powders he was buying – and nor would his customers, who bought the powders from the sites happykat.co.uk and perfectpowder.co.uk. Detheridge was jailed for three years, admitting he had been playing a cat-and-mouse game, trying to bring in legal drugs.

'He didn't know what effect they could have on the people he was supplying them to. They could have been substances which turned out to be lethal. He didn't know,' said Judge Adele Williams.[9]

Malcolm Bragg, criminal and financial investigation assistant director at the UK Border Agency, argues for stricter controls and more prohibition. 'Drugs devastate lives and communities and officers at the UK Border Agency, are determined to stop them reaching our streets.'[10] But a person with close links to the UK Border Agency said on condition of anonymity that the agency is overworked and under-resourced and that most single-kilo packages sent through UK airports by express courier service were more likely to pass through unimpeded than not.

The Serious Organised Crime Agency, responsible for targeting narcotraffickers, said in its 2012 report that it 'continued to disrupt the trade in new psychoactive substances. Work in conjunction with the Home Office and forensic providers is building a better picture of the range of psychoactive substances, both controlled and non-controlled, which are traded in the UK and this picture is continuously refreshed as new substances become available.'

It said that it had 'worked closely with the competent authorities in source countries such as China to seek to influence their response to the trade, including by encouraging them to tighten the legislation surrounding the substances. This produced some good results: China, for example, increased the controls of on-line sales and made mephedrone a controlled substance.'[11]

However, since mephedrone was banned in the UK and China in 2010, use has increased by twenty per cent and prices have doubled, and a torrent of new, legal drugs has emerged, each new ban prompting innovation from manufacturers keen to cash in on the legal highs craze. Banning a drug does not eliminate it from dealers' repertoires – instead, the range of drugs they offer simply expands. And in a classic piece of geographical displacement, whereby pressure to limit production in one country merely sees the trade shift to another nation, India now produces much of the UK's illegally imported mephedrone.

SOCA did, though, close down over 120 UK-based websites that continued to advertise controlled drugs,[12] but many of them simply reopened under different names, selling legal compounds.

It's highly likely that the flood of new chemicals will continue unabated. Alexander Shulgin has published a new book, *The Shulgin Index, Volume 1: Psychedelic Phenethylamines and Related Compounds*. It is the culmination of his life's work, and behind its sober, leather-bound cover lies more chemical data than most people could read in a lifetime. It covers over 1,300 compounds. It is, says its American publisher, 'an invaluable resource for researchers, physicians, chemists, and law enforcement'. It costs US$150, but its pages are available for free online, where volunteers are participating in a group edit and annotation. Volume 2, covering hundreds of new tryptamines, will be published within the next few years.[13]

British drug strategy was reviewed in May 2012, bringing together figures from the political establishment, harm reduction workers, scientists, teachers, frontline staff in hospital rehab units, celebrities, media commentators, entrepreneurs and some of the most senior police officers in the country. Members of the Home Affairs Select Committee (HASC), which is is appointed by the House of Commons to examine the policy, administration and expenditure of the Department of Health and its associated bodies, visited drug-producing countries such as Afghanistan and Colombia, and investigated the links between drugs and crime and disorder. It represented the highest-level mainstream political debate on drugs that has been seen in the UK for over a decade.

The home secretary, Theresa May, in a letter to the ACMD's Professor Les Iversen, acknowledged the thorniness of the problem:

> With the pace with which we are seeing new substances becoming available in the UK and, in addition to the more traditional routes of supply, with the internet playing a critical role by increasing the ways in which it is possible to buy NPS [novel psychoactive substances], the challenges to the Government, law enforcement and the forensic community are considerable.[14]

The announcement of the HASC's intention to open a debate around drugs policy was made at the same time that a series of influential international figures went public for the first time with their belief that the war on traditional drugs was not working, and that a new approach was needed. These included the presidents of Mexico, Guatemala and Colombia, all of whom have experienced events that, if replicated on the streets of Europe and the US, would be deemed unacceptable. Some 100,000 Mexicans have died in the last nine years as the country's criminal gangs supplied the US – the world's richest and greediest consumer of drugs – with cocaine, heroin, marijuana and methamphetamine.

Guatemala's president wrote in the *Guardian* that it was time for his country, as a key trans-shipment point for the drugs trade, to consider legalization. 'Guatemala will not fail to honour any of its international commitments to fighting drug trafficking. But nor are we willing to continue as dumb witnesses to a global self-deceit,' he wrote.[15]

Colombia's president, Juan Manuel Santos, similarly called for new thinking on the drugs trade in 2011: 'The world needs to discuss new approaches . . . we are basically still thinking within the same framework as we have done for the last forty years.' He went further, and said his country, the world's biggest producer of cocaine, would consider full legalization: 'If that means legalizing, and the world thinks that's the solution, I will welcome it. I'm not against it,' he told journalist John Mulholland.[16]

Public opinion has, in recent years, softened towards the idea of decriminalization of drugs – whose legal status I would contend has encouraged a growth in often-dangerous legal alternatives – or is at least more open to a debate around the topic. A few hardliners remain. I emailed Kathy Gyngell of the Centre for Policy Studies, who contributed to the HASC debate, to ask what her response to the increasingly dangerous legal highs market might be. She equated those buying and selling legal drugs on the internet with those who sexually abuse children. 'We do not stop pursuing crime or enforcing the law because crime persists or because the technology they use changes, whether the internet or not. If you took that view you might give up on tracking down paedophiles who have come to operate, as many criminals do, through the internet,' she said. However, the market in legal highs is currently, and rather self-evidently, legal in the UK, Europe and the US, and much of the rest of world. The sexual abuse of children by adults is not.

Gyngell offered commentary to the HASC together with boisterous right-wing media commentator Peter Hitchens. He cast the use of drugs, as the churchmen in 1920s America once did, in a moral light. 'I think that taking drugs is a wrong thing to do. I think there is a good reason for there being a law against it, and if people do it they should be punished according to the law. If we had held to that, then we would still have the levels of drug use which we had before the 1971 Act, which were minimal,' he said, citing no evidence whatsoever.

Hitchens declared that the rule of law, if properly and zealously applied, was in itself enough to dissuade young people from taking drugs: 'I think if you have a properly enforced law, where cannabis possession, which is illegal, is punished when detected, then one of the most important things you will do is you will armour people, who are under strong peer pressure from their school fellows to take drugs, against that . . . They can turn around and say, "No, I will not do that. I don't want to risk having a criminal record. I don't want to risk never being able to travel to the United States for the rest of my life. I don't think it's worth it."'

However, the hundreds of legal chemicals now available as replacements for cannabis are far more harmful than the illegal drug itself, as was noted by the drugs' inventor, John William Huffman. They may be smoked by any young person without fear of prosecution, or a criminal record, or the loss of travel rights, but they bring far more serious health consequences than marijuana. If what Hitchens and other prohibitionists say about peer pressure is true – that it can be dismissed with arguments around legality – then the smoking of these substances becomes infinitely more attractive to young first-time users.

Moderate commentators who sound a note of caution

over the availability of new drugs do exist. Oxford-educated consultant Kevin Sabet, who advocates more drug control, has advised three American presidents – Bill Clinton, George Bush and Barack Obama – on drug policy, and is a leading voice in the US on the topic. He was one of three main writers of President Obama's first National Drug Control Strategy, and he led the office's efforts on marijuana policy, legalization issues and emerging synthetic drug policy. Sabet agrees that legal highs present a major dilemma for law enforcement and legislators worldwide, but he says that simply outlawing these substances one by one is not a sustainable long-term strategy, and will not stop these drugs from being imported, produced or consumed. 'What needs to happen is the passing of legislation and laws that prohibit the sales, manufacture and consumption of whole classes of drugs – with the exception of drugs manufactured for medical or scientific purposes,' he told me.

He went on, 'I don't think legalization is the answer, since we know that making any drug legal lowers its price and increases its consumption and availability. We already have legal alcohol, tobacco and prescription drugs, and they are used at a much higher rate than illegal ones. Decriminalizing or legalizing these drugs would also increase their social acceptability. What we need is education combined with interventions that work to get users to stop using.'

How, though, can we explain from this perspective the drop in cigarette smoking witnessed in the US in the last sixty years? Today, 19.3 per cent of Americans smoke, but in the 1940s, it was around triple that figure.[17] Nicotine, as any smoker will tell you, is highly addictive and ruinously unhealthy. Yet public health information campaigns have massively reduced the prevalence of smoking in a generation, even in private spaces.

Frontline drug worker Mark Dunn works at the UK's first dedicated Club Drugs Clinic, part of the London Chelsea and Westminster Hospital. His branch of medicine treats what it terms 'problematic' users of club drugs, legal highs and other chemicals, whereas most drugs services focus on crack and heroin users. The clinic has been open eighteen months and staff have treated 250 patients, the majority of them gay men. Mephedrone is a major cause of concern for the clients Dunn sees, with some users now injecting the drug obsessively, chasing its short-lived and intensely euphoric high. 'Mephedrone is way, way up there; people are bingeing and having horrific come-downs and paranoia. They have to reinject and this causes serious damage to veins, along with abscesses. They become psychotic and unwell, some have been sectioned. Then they use again and they become unwell again very quickly. One patient had no previous psychiatric history, and he has been sectioned four times in the last year. It has destroyed his life,' he told me.

The last equally significant turning point in British drug culture, the rise of Ecstasy from 1988 onwards, was documented by journalist Sheryl Garrett in her 1999 book *Adventures in Wonderland*.[18] At the time, Garrett was the editor of the style, music and design magazine *The Face* and was an early adopter of MDMA. She now has mixed feelings about the way the drugs market has changed in recent years, and she cites bingeing as a phenomenon that has challenged her long-held beliefs over drug use. 'I was very pro-legalization and decriminalization, and used to cite arguments around historical prohibition. You know, "Banning drugs doesn't stop anyone taking them or making them and we're just funding organized crime, etc." Legalization would mean you would be able to buy pure drugs and that would be accompanied by a huge health information campaign. But the risk is that we are bingers in Britain. The first time I went to Ibiza the atmosphere was really relaxed, and people were taking one or two pills a night – not twenty. But that binge culture has also now spread worldwide. I see it everywhere I go now, and that scares me a bit.'

The binge culture also concerns Dr Adam Winstock, who as

well as working as a consultant in addictions, is also the creator and managing director of the Global Drug Survey, the world's largest research project involving users. 'Any discussion of de-criminalization, or even legalization, where products are sold with accurate labelling and dosage information sounds like a good idea, but proponents presume an adult population making informed decisions. I'm not sure the UK could handle that, as users here tend not to be restrained. Holland has a moderate and informed adult population. The UK does not do moderation,' he said.

Of course, it would be naïve to imagine that any moves towards a more liberal worldwide drug regime would be ushered in without complaint, however logical they might be. Matthew Collin, author of the 1998 book *Altered State*, the definitive history of Acid House and Ecstasy use in the UK and beyond,[19] says the debate around decriminalization demands political context. He believes realpolitik dictates that the legal situation in most countries is unlikely to be greatly transformed by developments in the research chemical market. 'A politician running for office is still open to attack from electoral enemies if he or she is seen to be "soft on drugs", while for a government in power, the rhetoric of the War on Drugs is still more easily comprehensible and reassuring, despite its cost, than a step into the unknown whose outcome might appear unpredictable and potentially frightening,' he said. 'Prohibition may not have worked, but in countries like the US in particular it's hard to see any political leader running on a legalization platform and having any chance of getting elected; the candidate would be ripped to pieces by attack advertisements during the campaign. And even if he or she made it into office, getting such a policy into the statute book would appear impossible. Look at the fierce resistance to Barack Obama's attempt to reform the American healthcare system – and then imagine if he was trying to convince Congress and the House of Representatives to legalize cocaine and heroin.'

Dealers who profit from the sale of banned drugs can have a clearer view of the problems of legalization than many other commentators. I conducted an in-depth interview with one of the most popular vendors of MDMA on the Silk Road, discussing the legalization of drugs across an encrypted email connection. 'The biggest issue I have with legalization is quantifying the pros and cons, what information do you base your decision on? Which metric is most important? Is it addiction rates, acute risk, economic cost, family breakdown, crime rates? It's easy to look at the gruesome prohibition-fueled civil war in Mexico, the private prison industry in the US, the gang-fighting over drugs that goes on in every city and draw the conclusion that legalization is the only humane and reasonable alternative, because all of those injustices are blatant and gruesome. It's harder to weigh the less apparent consequences, the subtle personal issues that easy access to drugs brings,' he said.

'As a dealer/vendor I get to see a much closer view of these problems, both in myself and others, and frankly it often upsets me. Many times I've had to stop selling to clients because they developed serious addiction issues. I know people who use MDMA every week and suffer serious memory and cognitive problems because of it; people who can't stop using coke despite not even enjoying it any more, people who have to pop Oxycodone just to make it through the day. Seeing it really wears me down. How many more people would there be like that if they could pop down to the convenience store and pick up an eight-ball of cocaine? Would they ultimately be better off if given access to whatever they wanted along with subsidized harm reduction and treatment programmes if needed? It's not an easy question to answer at all. I used to think that people should ultimately have agency over their own bodies and what they put in them, that the world was overwhelmingly worse off with prohibition than without it. I still feel that way, but over the past few years my view has become much more conflicted.'

In Europe, only Portugal has dared to experiment with radical moves towards decriminalization – of drugs far more harmful and addictive than the most popular recreational drugs whose effects many research chemicals and legal highs seek to emulate. Before 2001, heroin use in Portugal was rife. There were over 100,000 intravenous drug addicts in the country, and open-air drugs markets were commonplace. In 2001, the government decriminalized the possession and personal use of all drugs, including heroin and cocaine, and compelled users caught with banned substances to appear in front of special addiction panels, making drug use a health matter, rather than a crime.

'The changes that were made in Portugal provide an interesting before-and-after study on the possible effects of decriminalization,' the EMCDDA said.[20]

And indeed they do. In 2011, Joao Goulao, President of the Institute of Drugs and Drugs Addiction told journalists that the number of problem drug users had halved.[21] The rate of injectors also halved to about half a per cent of the population, below the levels seen in Britain and Italy. New HIV cases also dropped; in 2002, half of all new cases of the disease were injectors; today, that figure is 17.5 per cent.

Dr Adam Winstock says the UK missed a chance to institute a fresh approach to drugs policy when it banned mephedrone. 'The appearance of research chemicals like mephedrone in mainstream markets gave governments an opportunity to do something different. It's difficult for governments to retrospectively amend laws around existing substances, and mephedrone did offer a chance to use consumer or medicinal product regulations rather than the Misuse of Drugs Act. We missed that opportunity. Widespread use was initially reduced, but the drug migrated to street dealers, the price went up and many users thought that purity fell,' he told me.

Toxicologist John Ramsey says doing nothing is not an option. 'In terms of control, we have to do something in case some horrendous compound comes along. It's almost inevitable, just a matter of time. There is the potential for someone to drop dead the first time they take it. When you're tinkering with molecules you can get things wrong and there are surprises,' he warns.

But Ramsey is, like many since the legal high and research chemical market ballooned, conflicted over what to do now. Most feel legalization is also not an option, since many of the drugs are harmful, while leaving the industry unregulated is equally unsatisfactory. 'Why tolerate a lower standard of safety for recreational drugs than you do for pharmaceuticals?' Ramsey asked. 'The pharmaceutical argument is a risk/benefit one. If you have a tablet for headaches, you won't put up with one that has many side effects; it's easier to put up with the headache. If you have cancer, the chemotherapy will make you sick, impotent and will make your hair drop out, but you take it because if you don't, you'll die. The difficulty with recreational drugs is how do you make that call? The only positive effect of taking recreational drugs is pleasure, so how many side effects should we tolerate for a pleasurable drug?'

Other legal measures that may be considered in the UK are generic or analogue controls to update the Misuse of Drugs Act 1971. These kinds of laws seek to ban entire categories of drugs and have been in place in the US since 1986, to little effect. While the government's chief advisors have recommended further study in this area, experts at the Independent Scientific Committee on Drugs (ISCD) and the UK Drug Policy Commission argue that the courts could become filled with judicial claims against such a complex area of law. The ISCD said in a report citing recent admissions by the US that its own analogue laws were flawed:

> In everyday terminology, the term 'analogue' is often used to describe a substance which has major chemical structures in common with another chemical. To organic chemists, however, the term 'analogue' has a more precise

meaning. It is also the case that many chemicals that look alike and have similar chemical structures react very differently both in and out of the body. So the issue is not at all clear-cut.[22]

John Ramsey also believes that analogue laws would be impossible to enforce in the UK, and that other types of legislation would be equally difficult. Why, I asked him, could we not simply ban every chemical that had a psychoactive effect? 'The problem is collecting the evidence,' he said. 'When you are just appraising one or two compounds a year, that's achievable. When it's one a week, it isn't. Even if you did ban all the known ones, all of the possible substitutions, there would still be something you hadn't thought of. There are literally millions of organic chemicals. The issue is trying to find out which ones can be used as drugs and which ones can't. We don't have that knowledge.' He went on, 'The cannabinoid receptor agonists are a good example. They come from a wide variety of chemical groups, or families, so to try and define them chemically is extremely difficult. They all react with the CB1 (cannabinoid) receptor but there are some chemicals that are used in synthetic marijuana mixtures that are also used legitimately – there's one that is used as a lubricant in the manufacture of plastics, so it would have an impact on industry as well.'

What's more, an analogue law would leave people open to prosecution, said Ramsey, since they might believe they are selling a legal compound, only to have that assumption challenged in court, upon which they might face jail. We are only at the beginning of our understanding of how drugs work, he added, so writing laws on the basis of what they do is equally problematic. 'Not all drugs work because they have activity at a certain receptor site. For example, nitrous oxide, or laughing gas, has no known receptor reaction that laws could target. How, then, do we control that under analogue legislation? The science is much more complex than these simplistic models suggest,' he told me.

Danny Kushlick of campaigning UK NGO Transform argues that the best way to address the emergence of new drugs is not only to tackle drug laws in the round, but to ensure that law changes are in fact the last piece of the jigsaw puzzle. He is pragmatic, and notes that decriminalization of the regular market would not solve the problems of prohibition overnight. 'These aren't easy choices. You would have to incrementally introduce reforms and monitor their impact to assess what is working and what isn't,' he told me. 'Keep it open, transparent, democratic and open to science, and to critical review, impact assessment, value-for-money and cost-benefit analyses. If you keep applying those kind of tools and don't allow industry to run the show, you will put in place policies that are vastly better than the ones we have now and in terms of legal and illegal drugs,' Kushlick said.

But without deep change he believes we are doomed to repeat the errors of the past: 'You just can't develop sensible policy within the current paradigm. What we need to do is to step back and have a proper conversation among public NGOs, government departments and officials and develop a cohesive new set of policy principles. Then we can begin to form a genuinely coherent approach to managing the production, supply and use of the full range of psychoactive substances. And only *then* should you start introducing new legislation.'

Agnetha, the drug user who overdosed twice in a week on mislabelled but legal research chemicals, strikes a more combative note than Kushlick, saying: 'Oppression and prohibition don't work, they just drive people to accept ever higher risks by taking ever more dangerous chemicals, substituting each successively for the safer ones that were banned the day before. What we need are risk-aware and educated drug users, who can forego poisonous garbage and are able to satisfy their curiosity with comparatively safe chemicals. If the political elite can't quit prohibition, they should at least decriminalize all consumption and personal use and go after dealers and manufacturers only, and stop treating drugs indiscriminately.'

A frontline health worker dealing with heavy mephedrone users says he believes that mephedrone in particular has caused more harm to users since it was banned. 'It has now been cut with other substances and users are getting more health problems. Use has also grown since the ban. It's clear from the numbers of new drugs that are coming out that current legislation isn't having any impact on use at all. It's not helping clients at all. It isn't working. They are still using, still facing trouble, and it makes it harder to engage with them as it is now a criminal offence,' he told me.

Perhaps the difficulty we actually face in attempting to address the use and abuse of all drugs is in thinking there is actually a problem to solve at all. Humans have always used their wits and the products of their environment to change their states of consciousness. They have sought thrills and danger, adrenaline rushes or the comforting warmth of company and care, while the mystics and the masses have often sought precisely the same ends by different means. If the solution we seek is to eliminate danger, to end addiction, to prevent all negative consequences from drug use, then we are destined to fail as surely as we have done over the last century – especially when the web expands the chemical palette so dramatically.

We are currently unprepared legislatively, socioculturally, and practically for this, the next phase in the drugs market. Legalization is not the answer, banning drugs is not the answer, leaving things as they are – in complete unregulated anarchy in both the new and old drugs markets – is not the answer.

After many years observing this chemical underground, I have concluded that the changes the web has occasioned in the drug culture now mean legislators must act: there must be a concerted effort not only of harm reduction, but of urgent damage limitation. In scores of interviews, on thousands of forum posts, in dozens of forums, the explicit and implicit message from the drug users themselves is that no law will ever change nor ever has changed their desire to get high.

In New Zealand, lawmakers in August 2012 took an unprecedented step when associate health minister Peter Dunne announced innovative moves that would bring some state control over the country's uncontrolled legal highs industry. Approvals for new legal highs in the country, he said, would be granted once manufacturers had paid for scientific research into the substance's harm profiles. In New Zealand, the new drugs scene is focused on so-called 'party pills' – various piperazine mixes that emulate Ecstasy or amphetamines, and substitute marijuana compounds.

Those deemed to be low risk in clinical trials using humans and animals, estimated to cost two million New Zealand dollars (about one million pounds) for each new compound, would be allowed for general sale. 'We will no longer play the cat-and-mouse game of constantly chasing down substances after they are on the market,' Dunne told reporters.

What impact this would have on profitability for the firms producing the drugs is unclear, but the party pills industry in New Zealand is estimated to have made US$250 million in unregulated profits in the last decade.

Briefing papers by the government suggested that ten applications were expected in the first year, with a projection that one or two approvals would be given. This stricter regulatory approach looks likely to seize control of the uncontrolled legal highs and research chemicals market from the Chinese laboratories and profit-driven marketeers and entrepreneurs and hand it instead to the democratically elected government of New Zealand. A proactive and evidence-based harm reduction model such as this should, at a stroke, reduce the number of new drugs coming on the market in New Zealand. Furthermore, it depoliticizes the debate and delivers responsibility for regulating the trade to those best qualified to assess the undoubted harms some drugs can do: expert scientists and experienced doctors. It comes at little to no cost to the government, and users will be safer.

Acknowledging that these moves would create a legal synthetic drugs market, the world's first, Dunne told the *New Zealand Herald*: 'That is the absolute intention behind this regime. The problem in the past has been that we had a totally unregulated market with who knows what substances in these products. I am quite unapologetic about leading changes that will make things safer for young New Zealanders.'[23]

Professor David Nichols, the creator of some of the drugs that have killed users in this book, is unequivocal in his assertion that international drug laws are no longer fit for purpose. Many of the compounds his lab has produced, including 4–MTA, MDAI, 6–APB, 5–APB, 2C–B–FLY and bromo–dragonFLY, have been hijacked and sold on the international grey market as research chemicals or legal highs. These drugs, developed originally as part of the search for new medicines and to study neuropharmacology, have killed several young people in the last ten years. 'The first thing [research chemical retailers] do is to search for everything I have published,' he told me. 'You can get the papers for thirty or forty dollars. It's gotten bigger and become more widespread as a result of the internet. People might still have been interested [in the past], but being able to go to the internet and buy things and people communicating online so quickly has facilitated the development of this whole area. And it's only heading in one direction.'

For him, the nightmare scenario would be if some of the drugs he has manufactured escape the lab and are commercialized on an even larger scale than they have been hitherto. 'Say someone finds something with a psychostimulant and a hallucinogenic effect, and people try it and like it so they go and buy a few kilograms from China and buy a tableting machine – they're on eBay for a thousand dollars. Then imagine someone makes 50,000 tablets and distributes them, and people take them every weekend at raves and the cardiotoxicity is cumulative. What if the harm it does to your heart does not manifest on the first time you take it? People could take it, keep doing it, and all of a sudden after taking this for a month or two or thinking it's great stuff, they might find themselves going into hospital with heart problems, arrythmia, whatever.'

This, almost to the letter, is what happened with 4-MTA, though the health issue was more acute; 6-APB has similarly been commercialized for the research chemical market, though without such disastrous results – for the time being. Nichols says the solution to the serial tweaking of molecules and raiding of his work could be solved pretty easily. But it's a position that will win him no friends in high places. 'Legalize the safe ones. Mushrooms, mescaline and peyote, all have been used for thousands of years and have been shown to be safe. And marijuana, the most widely used drug in the US.' Nichols blames prohibition for the greater variety and strength of drugs available today. 'It is silly what they did with marijuana. We wouldn't have any of these synthetics, which are far more dangerous, if they had just said: "Marijuana has been used for thousands of years, it's an intoxicant, put it in the stores like alcohol, and make it so you have to be twenty-one to buy it." Regulate it in some way. We have beer, wine and liquor, and we could have different grades of marijuana like that. I had a grad student once and he said, and this may be axiomatic: "Make one drug illegal and another, more dangerous one will take its place."'

We must now allow drug users to make safer choices, and that means a gradual, tested, evaluated but concerted roll-back of all existing drug laws; particularly those concerning MDMA, marijuana, magic mushrooms and mescaline, for these are the drugs that most research chemicals seek to emulate. Only then will dangerous innovation end. Simultaneously, drug awareness classes should be compulsory at all schools with credible, evidenced and honest discussions of each drug's effects, good and bad, including alcohol and tobacco. This will not end the debate, or addiction, or reduce drug use. But it will mean those who choose to take drugs in the future will be better informed and safer, and the costs to society lower. Governments must now seize control of the market in new and old drugs from amateurs, criminals and gangsters.

Perhaps the web's final and most dramatic effect will be to strip drug culture of its mystique, its cachet of countercultural cool, to reveal that behind the magic and madness, there lie only molecules. At the end of it all, drugs are just carbon, hydrogen and a few other elements. They have their meaning projected onto them by users and the culture more widely. Remove the thrill of social transgression that acting illegally provides and reframe drug use in a clinical context, as a health issue, and that might change. We know in detail what the route we have taken for the last century results in: greater and more dangerous use. We now need a new approach and new data to analyse. It is not this book's argument that any drug is entirely safe; they demonstrably are not. But to persist in the digital age with this failed and arbitrary strategy of prohibition in the face of all the evidence that it increases harm is irresponsibly dangerous.

However, although some politicians are able to admit grudgingly to youthful experimentation with drugs, it seems few are willing to experiment even moderately with new approaches in policy now they have the power to effect positive change – even at a time when the people who vote for them are demanding exactly that, and when it is more urgent than ever before.

# 11

## *The End of the Road?*

In October 2012, five months after this book was first published, the question posed at the end of Chapter 9 was partially answered. But instead of a crossroads, we now stand at a fork in the road – and the alleged owner of the Silk Road site is in a jail cell, awaiting trial and imprisonment.

On 2 October 2013, Ross Ulbricht, a 29-year-old Texan man, was sitting (appropriately enough) in the science fiction section of the Glen Park branch of the San Francisco Public Library in San Francisco. Staff said they had seen him there a few times before, casually dressed and working on his laptop. His world was about to change just as dramatically as the Silk Road had revolutionized the drug market.

As Ulbricht was working, FBI agents entered the library and before he had a chance to shut his laptop down, they seized and arrested him. A few days earlier, Special Agent Christopher Tarbell of the FBI had sworn and signed written testimony before Frank Maas, US Magistrate Judge of the Southern District of New York, that Ulbricht was the Silk Road's Dread Pirate Roberts (DPR) – the site's founder and owner who took his buccaneering code name from *The Princess Bride* – and the arrest warrant had been issued. The FBI had built a comprehensive criminal case against Ulbricht, which they claimed prove his links with the Silk Road.

Officers later released screenshots of the laptop Ulbricht was using at the time of his arrest that showed he was logged into a chat client as 'Dread' – in conversation with an undercover agent – and logged onto the Silk Road under the username 'Mastermind', with access to, and total control of, the site's inner workings and its finances.[1] Legal documents show that Ulbricht now also stands formally accused of charges of ordering the murder of one man, and it is claimed he paid for the deaths of five more.[2] The FBI claimed it had unmasked the Dark Web's most infamous pirate outlaw. It had found his Bitcoin wallet containing 144,000 coins, worth US$20 million at the time of the arrest. Another wallet was seized on the site, containing 29,000 coins. The news sent the price of Bitcoin, which had been trading at around US$130, down to US$85 in less than three hours.

The criminal complaint[3] revealed much about the length of the investigation into Ulbricht, who, the FBI said, oversaw a US$1 billion enterprise – though it is worth noting that for most of the time the Silk Road was operating, bitcoins traded at far lower rates than the FBI calculated. The site was shuttered and an FBI seizure notice greeted visitors to the marketplace's homepage. All users' funds were seized, along with all vendors' escrow funds, and worldwide, people who had been using the site started to get arrested. The FBI could take satisfaction from a job well done.

At the time of writing, the trial of Ross Ulbricht has not taken place. Ulbricht denies all charges made against him and is defending his case. No assertion or assumption of his guilt or innocence is either made or implied here. Many more details and counterclaims will inevitably emerge, but what follows below are the FBI's allegations concerning an extraordinary tale of drugs, high technology and low cunning, taken from indictments, court filings and other legal papers.[4]

How did the FBI manage to find, locate and capture a man and an organization that cocked a snook at global drug laws so rudely – and for so long? If the FBI's version of events is

true, detectives simply started searching the web for the very
first mention of the Silk Road. Find that person, they figured,
and you'd have your man. On 27 January 2011 at 11.28 p.m.,
a poster at the Shroomery forums with the username Altoid
claimed that dubious honour. Altoid said:

> I came across this website called Silk Road. It's a Tor hidden
> service that claims to allow you to buy and sell anything
> online anonymously. I'm thinking of buying off it, but
> wanted to see if anyone here had heard of it and could
> recommend it. I found it through silkroad420wordpress.
> com, which, if you have a tor browser, directs you to
> the real site at http://tydgccykixpbu6uz.onion. Let me
> know what you think...[5]

The posting itself was made using Tor anonymizing software, so
police could not trace that account holder. A few days later, on
29 January 2011, at the Bitcointalk web forum a user named
Altoid posted to a thread titled 'A Heroin Store'. The thread was
a libertarian thought experiment where posters played with the
idea of selling banned drugs online, taking advantage of Bitcoin's
untraceability to assess its potential to dismantle state power over
individuals – a key aim for early adopters of the cryptocurrency.
At 7.44.51 p.m., Altoid wrote:

> What an awesome thread! You guys have a ton of great
> ideas. Has anyone seen Silk Road yet? It's kind of like an
> anonymous amazon.com. I don't think they have heroin
> on there, but they are selling other stuff. They basically use
> bitcoin and tor to broker anonymous transactions. It's at
> http://tydgccykixpbu6uz.onion. Those not familiar with
> Tor can go to silkroad420wordpress.com for instructions
> on how to access the .onion site. Let me know what you
> guys think.[6]

The subject matter, user name, and near-identical sign-off gave the police their first scent on the trail, it is claimed. The comment was later deleted by the user, but no matter – the word was out: Bitcoin, Tor and PGP encryption software meant drugs could be traded anonymously and with little chance of capture or surveillance, and then delivered via the regular postal system anywhere in the world. The tipping point came less than six months later, on 1 June 2011, at 4.20 p.m., when Adrian Chen of the news site Gawker posted a story titled 'The underground website where you can buy almost any drug'. This was the first mainstream mention of Silk Road.[7] The site grew rapidly and soon became the bustling marketplace described in chapter 9 – with police watching its every move.

In October 2011, Altoid popped back in to Bitcointalk – this time looking for a technical software developer.

> I'm looking for the best and brightest IT pro in the bitcoin community to be the lead developer in a venture-backed bitcoin startup company. Experience in a start-up environment is a plus, or just being super hard working, self-motivated, and creative. Compensation can be in the form of equity or a salary, or somewhere in-between.

And then, the clincher:

> If interested, please send your answers to the following questions to rossulbricht at gmail dot com.[8]

Police connected the Altoid identity with Ulbricht, and now had a strong digital and real-world trail to follow. Ulbricht's LinkedIn profile read:[9]

> I want to use economic theory as a means to abolish the use of coercion and aggression amongst mankind. Just

as slavery has been abolished most everywhere, I believe violence, coercion and all forms of force by one person over another can come to an end. The most widespread and systemic use of force is amongst institutions and governments, so this is my current point of effort. The best way to change a government is to change the minds of the governed, however. To that end, I am creating an economic simulation to give people a first-hand experience of what it would be like to live in a world without the systemic use of force.

This 'economic simulation' the FBI argues, was Silk Road.

DPR, unusually for a Dark Web drug market kingpin, had a book club on his forums, where he directed interested readers to works by US libertarian theorist and historian Murray Rothbard and the agorist thinker Samuel Konkin. Agorists reject state power and control and believe in 'counter-economic acts', such as drug dealing or other grey-market activities that withdraw individual support for the state which normally takes the form of payment of taxes, framing such acts as a means to empower the individual and overcome government oppression. DPR, whose forum signature file often carried links to the mises.org site – also dedicated to libertarian thinking – left the following post on the forum on 10 March 2012:

Rothbard and Konkin were the two main inspirations for creating Silk Road, so it is interesting to see that they disagreed on some things. They both have the same ends: abolition of the state.

Silk Road, in this reading, was not just an attempt to buy and sell drugs and get high, it was the first step in creating an anti-state consciousness, leveraging drug users' dissent for a much larger ideological goal.

That claim was made explicit by DPR in a forum posting on 18 September 2013:

I'm a drug user. I've smoked pot, I've tripped on shrooms, MDMA, and others. I never hurt a living soul any time I did either. If anything I was even more loving, empathetic, creative, bold, thoughtful, etc. than when sober.

How many people have starved to death because the agents of the state use up all of the resources to fight end-less wars, subsidize invasive and destructive projects and do their best to keep everyone else under their thumbs? There is nothing redeeming there. The little good they do eek [*sic*] out is just enough to get elected and avoid revolt.

In short, Silk Road is an example of a moral culture where peace, cooperation and ethical competition are the norm, and violence and fraud are found only on the margin. This is opposed to the nation states of the world, where violence and fraud are used as a means to control their citizenry and to dominate one another. The people that make up human society at large need to adopt the Silk Road credo or we may all perish, or live under the thumbs of tyrants indefinitely.

A search for Ulbricht's name on Amazon.com returns a thesis he wrote in 2009, 'Europium Oxide Thin-Films: Exploration of Epitaxy and Strain'.[10] 'This manuscript is the culmination of my career as a scientist,' Ulbricht wrote in his notes. 'It marks the fulfillment of my goal to expand the boundaries of human understanding of nature. The rest of my life is dedicated to bringing freedom from oppression to the people of the world.'

Speaking to a friend in an interview carried out on 6 December 2012, which they videoed and posted online,[11] Ulbricht was asked what he wanted to achieve in the next

20 years. He responded, 'I want to have had a positive and substantial impact on humanity by that time.'

Ulbricht had no idea he was facing a vast team of intelligence investigators dedicated to his capture at any cost, or that he had already played a part in his own downfall. But right from the outset, Ulbricht, if he is indeed DPR, had failed to observe even the most basic principle of OPSEC – namely, operational security, as practised by hackers. Good OPSEC means covering your tracks at every stage of an operation, mitigating risk at all times, and always being one step ahead of any adversary.

In the weeks following Ulbricht's arrest, 'thegrugq', a renowned, pseudonymous, Bangkok-based IT security professional, discussed the case with me by email, offering a fascinating analysis of the Silk Road story from the perspective of security and espionage – which is what the case most closely resembles:

> While I personally don't advocate drug use, I don't believe that prohibition is the correct policy. It has been demonstrated empirically over the last 100 years that prohibition is a failed policy.
>
> Prohibition places control over the production and distribution of potentially dangerous substances into the hands of extralegal groups. I am very much against creating well-funded, violently inclined extralegal groups. I am against criminalising and demonising the sick and needy. It strikes me as extremely counterproductive.
>
> I am neither pro-drug or pro-underground market. I study it from an objective and neutral position to understand how adversaries interact in an environment. I look at it like chess, or any other competitive game. There are objectives, constraints, strategies and tactics. Both sides play to win, but you don't champion one side over the other.

So to me Silk Road falls within that 'great game' space;
what are the objectives and constraints? Which strategies
would be most effective in this operational environment?
Which tactics?

If the FBI's allegations are true, thegrugq says, Ulbricht's errors
were due to poor 'compartmentation' of his identities.

Once one identity was compromised, all the others were
exposed because they were publicly linked. He failed
to keep separate the identities used for: marketing Silk
Road, doing human resourcing, learning the technology,
and pursuing his personal interests. Once one identity
was compromised, all the others were exposed because
they were publicly linked.

On 5 March 2012 a new user, 'frosty', registered at Stackoverflow,
a Q&A website for coders facing trouble with their site or other
IT project. Frosty registered using Ross Ulbricht's gmail address.
On 16 March 2012, Frosty asked: 'How can I connect to a Tor
hidden service using curl in PHP?'

The FBI alleges that, a minute after posting the question,
Ulbricht changed the account's email to a fake account, frosty@
frosty.com. That same fake address would later be found on the
Silk Road server, where police identified code 'identical to lines
of code quoted in [that] posting'.

The site was growing, and the net was closing.

The FBI began to infiltrate the Silk Road discussion forums
where buyers and sellers were chatting about quality and service,
making requests for new products and sharing gossip. The forums'
strength – their sense of community and relative openness – had
now become a vulnerability. The FBI then infiltrated the Silk
Road market itself, making over 100 test purchases – most of
which turned out to be drugs of high purity.

The FBI says that its next step was to lure DPR into a trap. In December 2012, an undercover officer posing as a major dealer contacted DPR offering to sell bulk quantities of cocaine on the site. DPR's trusted employee Curtis Clark Green, a 47–year–old man from Utah who had worked for DPR on the site under the alias 'chronicpain',[12] became the middle man in the deal and in January 2013 accepted delivery of 1 kilo of cocaine. He was arrested, and charged with the drug offence. Meanwhile, DPR contacted the drug dealer who supplied the kilogram of cocaine and informed him that Green had been arrested – and that he had stolen other users' funds. DPR asked the drug dealer – in reality an undercover agent – to torture Green and force him to return the stolen bitcoins. Then he had a change of heart, and upgraded the torture instruction to murder, since Green knew too much and would be sure to cooperate with police. The FBI officers who had posed as delivery agents forced Green to turn informer – an easy task once they showed him proof that DPR wanted him dead. The FBI staged his murder in February 2013, mocking up photographs of the supposed victim.[13] Ulbricht was charged with murder–for–hire in Maryland in October 2013 following his arrest.

This would not be the only murder he would commission, said investigators. Court documents claim that on 13 March 2013 a user named FriendlyChemist messaged DPR threatening to reveal the names and addresses of thousands of Silk Road users and dealers.

FriendlyChemist said he was in trouble. He had accrued debts with his own supplier, and demanded US$500,000 from DPR to stay silent, later sending examples of the data he would release to prove he was telling the truth. In the FBI's version of events, a convoluted and bizarre extortion and blackmail attempt followed. DPR contacted his blackmailers' suppliers, who operated under the alias redandwhite (a common nickname for the feared Hell's Angels motorcycle gang), to work out a deal, and invited them on to the site to vend. DPR then decided to ask redandwhite to kill FriendlyChemist:

> I would like to put a bounty on his head if it's not too much trouble for you. What would be an adequate amount of money to motivate you to find him? Necessities like this do happen from time to time for a person in my position.

The FBI claims that redandwhite readily accepted a fee of US$150,000 in Bitcoin to kill a man who owed him US$500,000.

Investigators say the extortion ended in a payment by DPR to redandwhite of 1,670 Bitcoin, along with address details of a man living in White Rock, British Columbia in March 2013. Proof that the contract had been fulfilled was provided by redandwhite in the form of photos mailed to DPR, who deleted them. Thus redandwhite became a trusted confidant for DPR – so trusted, that it is alleged DPR later arranged the purchase of fake IDs from him.

Things got stranger yet. On 18 March 2013, the hacktivist known only as 'th3j35t3r' posted the following cryptic tweet, along with a bogus link that attempted to inject code into the browsers of users who clicked on it: 'I dun told ya. Dread. Pirate. Roberts. >> http://goo.gl/gkhpC << stay frosty.'

th3j35t3r is a conservative hacker who has in the past targeted such groups as Anonymous, the radical hacking collective, and Silk Road would have been an obvious target for his attention. The sign–off 'stay frosty' could be a simple slang usage meaning 'stay cool' – or it could have been an intimation that he had hacked the Silk Road servers and there found mention of this fake address, connected with Ulbricht.

Prosecutors alleged a further four conspiracy-to-murder cases. Ulbricht was accused of paying redandwhite to kill again – in April 2013 – this time targeting a dealer, tony76, who had scammed many customers on Silk Road. Upon hearing that tony76 worked in a house with three associates, DPR decided to have them all killed at once, this time for US$500,000 in Bitcoin. The blockchain, though anonymous, is public and traceable and this payment was confirmed.

Over the course of several months, the FBI continued to lean on key informants, wringing information out of them, and even managing to turn one of the site's top 1% of dealers – the heroin, cocaine and methamphetamine dealer known as Nod. Nod was raided in July 2013 and began cooperating with investigators, turning over his account to officers who set up customers.

The search for Ulbricht was starting to accelerate. On 10 July 2013, Canadian border police intercepted a package during, they say, a routine border search. It contained nine fake IDS, each in a different name yet all carrying a photograph of the same person: Ulbricht. It was addressed to his apartment. On 23 July 2013, in a part of the FBI investigation that the agency has kept entirely secret, the Silk Road servers were found – one of them in Iceland – and copied, so it is certain that their archives contain information that will result in many more arrests in the future. On 26 July 2013, agents visited Ulbricht at home, making inquiries about the false IDs. He did not answer questions, but instead quoted the Silk Road's market instructions addressed to nervous users: 'Anyone can post anyone anything in the post.'

Rather than running with his fortune, or deleting his hard drive, or closing the site, or moving the servers, for reasons that no one can yet understand or explain, investigators say Ulbricht continued living in his rented room. The FBI says he kept records of all of his wrongdoing, with dates, and an asset list that included Silk Road at US$104 million.[14] On 2 October 2013, the FBI finally made its arrest. In December 2013, Ulbricht was refused bail on the basis that he posed a flight danger.[15] His trial is scheduled for November 2014.

After a two-and-a-half-year operation to capture Ulbricht and close down the Silk Road, the site remained shut for just four weeks. In November 2013, it came back online with new support staff, a new Dread Pirate Roberts, and new forums. It looked exactly the same, but it is certain that behind the scenes many things were not as they seemed. A goading new landing page welcomed users to the site, mocking the FBI's takedown notice by overtyping the feds' message that the market had been seized with the phrase 'has risen again'.

The new DPR seemed much more technically competent than the previous owner, dropping cryptographic hints at a hacker sensibility, and from the outset seemed determined to build a reputation as a compassionate, generous outlaw, setting up a Bitcoin wallet for donations to a disaster relief fund to the Philippines, which had suffered a devastating typhoon that same month. In an interview he gave to me that I published at medium.com in October 2013,[16] the new DPR disavowed the tactics both of his alleged predecessor and of the war on drugs: 'Ultimately you cannot stop people doing drugs, but you can make it safer for them, and get people off the streets and away from violence – which is what we stand for.'

The new Silk Road market ran for a few weeks in an atmosphere of tense disbelief, and many of the thousands of vendors from its first iteration stayed away.

On 20 December, Reuters reported that Ulbricht's lawyer, Josh Dratel of New York, best known for his work with terror suspects and Guantanamo detainees, was discussing a possible plea deal.[17] The same day, police arrested three other key figures[18] involved in running the old Silk Road forums, and it looked once more like the site would be closed – perhaps this time permanently. A chaotic few weeks ensued as the site's owners quickly and deliberately dismantled their online personae and, supposedly, new staff were recruited, but trust and credibility was quickly being lost as the servers were constantly down.

In February 2015, the owners of Silk Road 2.0 claimed the site had been hacked, and all users' and vendors' coins – whose value was estimated between $2.6m and $40m, were stolen. The marketplace and forums descended into farcical, conspiratorial scenes, but the site remained open, and some dealers and users continued to trade directly without using escrow services, where coins are stored in a central, vulnerable wallet.

It would not have made very much difference if the Silk Road 2.0 had closed, for there are now dozens of different sites from which to anonymously buy illegal drugs online with Bitcoin, and information about them is spread all over the web, especially on reddit.com.[19] Trust was eroded, and the model of centralised marketplaces was severely damaged by the events of early 2014, but the concept – of buying drugs online anonymously with bitcoins via the mail – endures and will surely develop in the coming years.

In the same way that the research chemical scene went truly mainstream only when mephedrone was banned, busting the Silk Road has simply scattered the seeds of digital dissent and chemical disobedience far and wide. Setting up a market can now be done in a day or so. Whether users trust it or not remains to be seen – and if recent history is anything to go by, trust is the last thing anyone on the Dark Web should do.

A whole new sector of organized criminals is now involved in the online drug market – but police are hot on their tails.

Throughout 2013, users shopped at Agora, Blue Sky, Flomarket, Tormarket, Modern Culture, Budster, Pandora, RoadSilk, Outlaw, White Rabbit, Three Hares Bazaar, Black Services Market, Tortuga, Litebay, Deepbay, Freebay, Ramp Vault43 and Drugslist, along with Sheep and longstanding player Black Market Reloaded (BMR). Many of these sites were straight-up scams, but several played a longer con. Sheep fleeced its customers in a scam that saw the site's owners abscond in December 2013 with around US$40 million of its users' coins. Others, including

BMR, were run by honest criminals simply making money from prohibition. BMR closed in an orderly fashion after allowing all users to withdraw their money from the on-site wallets. It appears that the site admin for BMR, known online as backopy, was the only Dark Web market operator during this period who did not steal his users' money, order assassinations, cooperate with police – or get caught. In fact, when 200 bitcoins were hacked from his vendors' accounts, he repaid them from his personal stash in a move that cost him around US$16,000. In February 2014, he opened a new site, called Utopia. That site, too, was closed after just a few days in February when Dutch police mounted a sting operation and arrested five men on drugs and weapons charges.

'What I've seen so far is that a lot of the "next gen" guys are not technologists (because, obviously, if you are a skilled technologist you make money by playing the startup lottery, not risking jailtime),' says thegrugq. 'They are using rent-a-coder outsourcing sites to get an unwitting third-party developer to build the site for them. They are compartmenting themselves from the infrastructure component of the business and presumably, focusing on the capitalization, marketing and so on.'

As well as Tor, which the NSA has been unable to defeat despite its best efforts, new anonymity networks have gained popularity, including i2p, where a new trading platform named simply 'The marketplace' innovated with multi-signature Bitcoin transactions, doing away with the need for trust in a third party holding funds in escrow. New, secure communication systems such as bitmessage and cryptocat[20] came on stream, and all of the software was open source, negating attempts by the National Security Agency (NSA) to bully manufacturers into inserting malicious backdoors or undermining encryption.

After the sharp crash in value following Ulbricht's arrest and the closure of the original Silk Road, Bitcoin soared to unprecedented heights: an 8,723% increase in 2013.[21] This made millionaires out of early adopters and any drug dealer canny enough to have hoarded his coins rather than cashing out. On 29 November 2013, a single bitcoin was worth more than an ounce of gold.[22] Though this rise in value was attributed in part to increasing adoption of the currency in China, where investors used it to move capital out of the country and avoid capital controls, and to a positive statement by the US Federal reserve on virtual currencies in late 2013, there is no denying that the publicity generated by the Silk Road case enhanced its popularity too. In January 2014, Bitcoin was trading north of US$1,000, meaning the value of Ulbricht's seized wallet has rocketed from US$20 million to US$144 million.

The Silk Road bust brought the currency – and the concept of Dark Web drug markets – to much greater prominence, and promoted the idea of an anonymous, uncontrollable, digital means of value transmission far beyond its original constituency comprised of drug users and hackers. While the market model does now look compromised, direct trading relationships based on trust, with identities verified by vendors' use of cryptographic keys and PGP-signed messages, are gaining popularity. The ingenuity of this technologically adept avant garde will continue to evolve and adapt to whatever internal and external threats it faces.

Having already revolutionized the online drug trade, Bitcoin's ability to disrupt and transform other industries, both illicit and lawful, is only just beginning to make itself felt. And the early users of the cryptographic technology that underpins Bitcoin, Tor and encrypted email today look remarkably forward-thinking, where once they looked cautious, if not downright paranoid.

This story would be incomplete without acknowledging the events of summer 2013. In May of that year, another 29-year-old American, Edward Snowden, a contractor at the NSA in the US, revealed that the US and UK governments were targeting the actions of every single net and telephone user completely indiscriminately. Snowden fled the US and delivered to *Guardian* journalist Green Greenwald a cache of thousands of documents that detailed the inner workings of the spy agency. In June, Snowden revealed himself as the source and confirmed that US and UK agents have every user on the internet – not just adversaries – under permanent, indiscriminate dragnet digital surveillance – just as predicted by John Callas in Chapter 9.

The FBI's capture of Ross Ulbricht may well have relied on old-fashioned detective methods, and the court case may determine the truth of its assertions, but it is not unprecedented for the FBI to create facts after the event, in a process known as 'parallel construction'. This is the legal equivalent of keeping two sets of company accounts – a kind of 'information laundering' that denies the accused a fair trial under US law.

Reuters revealed in August 2013 the existence of a Special Operations Division (SOD) within the Drug Enforcement Agency (DEA) that receives and distributes tips gathered by the NSA. Using that information, the DEA makes arrests, and hides the source of that information.[23]

Snowden's cache of documents revealed the existence of agency-run programs with names such as Prism, Tempora and XKeyscore, as well as failed attempts to break the Tor network and the encryption used by banks, shops and businesses. The Xkeyscore program allows analysts in the US to learn the IP address of every person who visits any website. Prism boasted of direct access to servers of a number of firms, including Google, Apple and Facebook. Tempora, which was managed by the NSA's British counterpart Government Communications Headquarters (GCHQ), went further: it hijacked undersea fibre optic internet cables and copied all of the information – every phone call, SMS, email, web page request or other digital act that passed through them. It then shared that information with the US, enabling it to circumvent constitutional law. It is possible these methods were used, in some way, to target the Silk Road.

Following his revelations, Snowden was chased from Hong Kong to Russia by US agents, who considered him a criminal. In June, he spent weeks living in Sheremetyevo Airport, stateless, with his passport revoked by his home nation. That summer, the US even grounded the plane of Bolivian president Evo Morales in its hunt for Snowden, and closed European airspace in a failed attempt to capture him. Snowden's laptop, seen in occasional press conferences, carries just two stickers, one showing the Tor browser icon, and the other the name of the Electronic Frontier Foundation, which was set up by John Gilmore, who also founded the rec.drugs newsgroups mentioned in chapter 3.

Hanni Fakhoury, a lawyer at the Electronic Frontier Foundation, told me by email that if certain NSA surveillance tools were used in the capture of Ulbricht, it could have an impact on the trial:

> If one of the [NSA's] bulk collection programs was used and evidence was then sent to DEA or FBI or any other domestic law enforcement agency investigating the case, or any other form of unannounced electronic surveillance was used, then EFF believes that fact must be disclosed because it could be relevant to raising a legal challenge to how evidence was collected.'

Drugs, technology and personal freedom are now more closely connected than ever before, and the strands of this book continue to entwine: technology and drug culture inspire attack and counter-attack, driving each other to ever greater innovation.

What is clear from the Silk Road story of 2011–2013 is that the phenomenon of online sourcing of illegal drugs has only just begun. The journalist Gwern Branwen, who has documented the Silk Road in detail,[24] told me by email that he believes the future of Dark Web drug markets looks strong:

Now that the business model has been proven beyond a doubt with audited profitability figures (you can thank the FBI for that one), every geek in the world understands that they can become a millionaire if they dare. It's back to whack-a-mole: new markets will pop up, and will run until they get hammered down or rip-and-run. Evolution means the ones who leak their identities like Silk Road or Sheep, or who write bad code... will either fix their problems or get weeded out and replaced.

thegrugq sounds a note of caution to those who might trust the Tor anonymizing software with decisions that might one day cost them their freedom:

> I, personally, don't believe that Tor is safe when used to operate an online marketplace. It was not designed to protect against a global passive adversary, that is: someone who can see all the internet traffic everywhere. It turns outs that the NSA is such an adversary. I am even suspicious that better alternatives haven't received any funding while Tor has received millions. It seems like the Feds are feeding a weed they're not afraid of to grow and strangle the alternative options. The security that Tor provides works best at circumventing censorship. To provide the security necessary for operating against a nation state-level adversary requires significant behavioural and operational changes as well, such as a compartmented anonymous laptop and internet connection. Tor alone isn't sufficiently safe against a nation state-level adversary. Ross Ulbricht had a nation state-level adversary.

Sites such as The Farmers' Market, mentioned in the previous chapter, were known to just a few a few hundred, or perhaps a thousand, people worldwide, while other now-defunct forums and sites such as Sandoz Labs, The Bible, Follow the White Rabbit, Follow The Green Biker, Su.pplier.info, Lanel, Open Sore, Club Silencio, The Drugstor, A Figment of Your Imagination, The Looking Glass and Binary Blue Stars were known to just a few hundred people in total. Open Source, also known as the Open Vendor Database, marked the moment when open sourcing of illegal drugs started gathering momentum – as its names suggest, the business was carried out in the open, and it gathered a few thousand users. Silk Road was the upgrade, brokering nearly two million deals in less than three years and having hundreds of thousands of users.

It is unlikely that online dealing will ever replace conventional trading completely, since many drug purchases are made on impulse. But it is also untrue that most people who use drugs are addicts, or unable to control their impulses. Many simply want to purchase high-quality, unadulterated drugs, to make their own decisions as to what happens in their own bodies, and to weigh the risks to their health as autonomous, rational adults. Dark Web drug markets are a perfect interim solution for these users. Drug dealing and, by extension, use, is now so massively distributed that it is no longer feasible to pursue every user, yet governments persist, as if the internet had changed nothing. The disruptive energy of the net and many of its most technologically advanced citizens is still, more than 40 years since the first online drug deal, being channeled into acts of digital and chemical dissent.

An anonymous user at reddit.com, who documented many of the sites just mentioned, said:

When I first started in this shit, it was about getting high. Later on, it was about getting high and collecting sources. Later on it was for the community and getting high as well. Later on I did make a decent bit of cash but nothing that was worth the risk in retrospect... But for the last several years I have been involved in this shit, it isn't about me getting sources or about me getting high or about me making any money at all, it is 100% about telling the government to fuck off, stop enslaving us, we will not be defeated by you and we will fight and we will risk going to prison and dying in prison to stand up against your lies...[25]

Such political vitriol may sound untempered, and the Silk Road's claims of libertarianism and freedom-fighting can at times seem to be the far-fetched and self-serving delusions of politically naïve youngsters, but consider instead how the world looks to anyone whose friends have been jailed for years and denied an education or opportunity to travel as punishment for a crime that is as victimless, in their view, as drug-taking.

A poster named Nightcrawler on Silk Road forums in 2013 told me that he was on the site not to buy and sell drugs, but instead to help people understand the limits on their privacy in the digital age, and how to protect themselves against state intrusion. He offered an interesting analysis of current political realities:

I am a child of the Cold War. My father and uncles donned uniforms when war was declared in 1939. Both of my spouse's parents were in uniform as well. They went over to Europe to fight the type of fascist crap we are now told is necessary to 'protect us' or 'protect the children'.

When I was growing up, like so many others of my generation, I was told, repeatedly, that we were 'free' and that the authorities trawling through people's reading habits at the library only happened in Soviet Russia. I well remember that, even as late as the 1980s, Nicolae Ceauşescu was excoriated in the Western press because his regime had a handful of cameras in the streets surveilling the population of Bucharest. One is led to wonder what he would have made of modern Britain with, what, 4 million cameras?

In January 2014, *Guardian* journalist John Harris offered a similar, sobering perspective:

> Consider what the essential functions of the modern state look like to any politicised person under 30. The state comes to the rescue of banks while snatching away benefits. It strides into sovereign countries, and commits serial human rights abuses. It subjects doctors, nurses and teachers to ludicrous targets. It watches us constantly via CCTV, and hacks our email and phone data. It farms out some of its dirtiest business to private firms... [C]ontrary to the vanities of the 'free market', neoliberal capitalism needs the big centralised state to clear its way and enforce its insanities.[26]

The immeasurable threats of terrorists, paedophiles, drug-traffickers and hackers are now being used by the state to justify complete dragnet surveillance of all net users – which, in 2014, means most people in the developed world.

Each year, thousands of agency operatives funded by millions of taxpayers' dollars and pounds are now being poured into the permanent scrutiny of the web and the Dark Web, creating a digital panopticon that criminalizes drug users as effectively as it stifles democratic debate. Ultimately, the digital dissidents argue, the war on drugs is not a war on hedonism, on chemistry or on crime; it is a war on individual freedom and privacy. They say that any government clampdown on encryption, Bitcoin, Tor or PGP in coming years must be seen for what it is: an attempt by the state to continue monitoring the communications, interests, relationships and financial transactions of all its citizens. Prohibition will make a convenient cover story.

thegrugq offers his own rejection of dragnet surveillance and prohibition in personal terms:

> Prohibition is the only policy that has been so thoroughly disproved and continues to be enforced. I don't believe that any psychoactive drugs should be 'illegal', but rather that access [to them] should be controlled. I fully support harm reduction. I just lost a close friend, Barnaby Jack,[27] to a drug overdose. It was accidental, but maybe he would be alive if he knew the potency and purity of the substances he was abusing.
>
> Prohibition in no way stopped him from dying, if anything it made it more likely that he would end up dead. The policy limited his options for seeking help, it creates a stigma, limits people's ability to get gainful employment, it encourages them to engage in unsafe practices such as taking drugs in secret and alone, where they have no access to help should something go wrong. There are valid reasons for an intelligence agency to want to 'collect all the things', but drugs have no place in that discussion.

The real danger of the new digital war on privacy, says John Callas, co-creator of PGP and Silent Circle, an encrypted communications service provider, is to democracy itself. He told me:

> The biggest threat comes from secret programs backed by secret laws with secret courts overseeing them and issuing secret decisions interpreted in secret... by the staff lawyers of the people who do the secret programs. That's not the way a free society operates.
>
> The next biggest threat comes from the erosion of free discourse that happens when we have surveillance. It's well documented that people just start playing it safe. After that, we have the fact that programs always get

mission creep. Laws for exceptional circumstances get used in general. Mechanisms used to fight terrorism this decade will be used on political opponents in a generation. After all, the opposing party wants nothing more than to destroy our way of life, right? The fixes to this are not technical. They can be supported by technology, but they are social and legal.

The solutions Callas describes may take years to develop, if they are ever developed at all. A new, digital-era social and personal consciousness has long been dreamed of, but never realized – one that harnesses the web's connective powers to create a progressive, peaceful international consensus, rather than the solipsism and narcissism of the kind enabled by social media.

The legal fixes look more likely to come initially from the US, with its constitutional guarantees of freedom. As this book goes to press, President Obama has announced a reform of the NSA that may see its powers limited, though it is perhaps optimistic in the extreme to think any progress made will be permanent, American politicians of every hue are broadly united in their belief that democracy is more important than permanent surveillance of the entire web and everyone on it. And in July 2013, the EU announced that its parliamentary Civil Liberties Committee would mount an 'in-depth inquiry into the US surveillance programmes, including the bugging of EU premises'.[28]

In the UK, coalition leader David Cameron announced no such moves to investigate the illegal activities of GCHQ; instead, he has created a web-access policy that severely limits free speech on topics as diverse as sexuality, drug use and politics, in the guise of limiting children's exposure to danger online.

*Wired* magazine reported:

> As well as pornography, users may automatically be opted in to blocks on 'violent material', 'extremist related

content', 'anorexia and eating disorder websites' and 'suicide related websites', 'alcohol' and 'smoking'. But the list doesn't stop there. It even extends to blocking 'web forums' and 'esoteric material'.[29]

The twenty-first century will be shaped, technologically and politically, by some of the forces documented in this book. Every time a Dark Web market – of any kind – is closed by the state, it will demonstrate only that running one is immensely profitable. But the future of personal drug markets lies in the hands of the technically adept young people who are already devising new, decentralized marketplaces that cannot be attacked unless governments begin X-raying and scanning every piece of mail sent and received in the entire world.

Perhaps such inexcusable state intrusion into the physical, rather than the digital, world would at last wake many of us from our complacent slumber.

# Epilogue

On 30 January 2009, the year following the safrole burn that caused the quality and quantity of MDMA to plummet globally, there was a minor meeting in the margins of larger trade talks in Brussels between the EU Commission and the Chinese government. Commission President José Manuel Barroso, a staunch prohibitionist, and Chinese Prime Minister Wen Jiabao signed an agreement on drug precursors and other chemicals used in the illicit drugs manufacturing industry. The objective of the 2009 bilateral agreement between China and the EU was to monitor the trafficking of precursors and to prevent their diversion from legitimate trades. The agreement became active on 11 July 2009, and was the first time that China and the EU established a system of monitoring the legal movements of precursors.[1]

From the banned chemical para-methyl-ketone or PMK, there lie only a short series of reactions to MDMA, just as there are from safrole. Chemicals such as these cost mere dollars per kilo and produce drugs that are worth tens of thousands; a move at which any medieval alchemist would surely marvel. Even the production of naturally derived drugs such as cocaine demands potassium permanganate, while a key reactant in the conversion of opium to heroin is acetic anhydride. But all these chemicals have legitimate uses in various industrial sectors from perfumery and flavouring to medicines and water treatment, so cannot simply be banned.

The main initial effect of the bilateral agreement was to entrench the MDMA drought. But suddenly in 2011, the quality and quantity of MDMA mysteriously shot up all over Europe. It was obvious that something was happening. Researcher Peter van Dijk of the Trimbos Institute in the Netherlands confirmed the news to the Associated Press: 'There's a large amount of pills going around containing a high dosage.' A full twenty-five per cent of all pills had a higher-than-average dose, at 140 mg. '[They] are too potent for many people,' said Van Dijk.[2] Test centres in Amsterdam found pills containing large and potentially dangerous doses of MDMA – whereas normally there might be 80–145 mg in each pill, the quantity in the new Dutch pills was racing up to 200, even 220 mg. There were a number of overdoses, some of them fatal, as users who were accustomed to taking several pills in a night found themselves dangerously overheating.

At an event in London's Alexandra Palace in late 2011, two young men died over the same weekend: twenty-one-year-old Lloyd Jones of Oswestry, Shropshire, died after midnight on 27 November, and twenty-year-old Richard Baker died at 7.29 a.m. that day. Initial rumours suggesting that they had taken only Ecstasy were met with disbelief. But coroner's reports in July 2012 confirmed that they had only MDMA in their systems.

Immediately British newspapers claimed that a new, superstrength Chinese variant of MDMA was responsible. They were almost right. The sudden upswing in quality and quantity was actually due to innovative chemists in the Netherlands, who had discovered a novel solution to the international control of precursors for MDMA. The organized crime syndicates behind the drugs' production had simply shifted from the now more closely controlled precursor liquid PMK to solid PMK-glycidate, a legal analogue of the banned precursor, imported from China.

'The process is called "masking" and the increasing use of masked precursors presents myriad challenges to drug control authorities', said Justice Tettey, chief of the Laboratory and Scientific Section of the UNODC. 'Traffickers can change the appearance of a chemical by "tweaking" the chemistry to circumvent controls, and easily recover the parent compound prior to use in illicit manufacture. The possibilities are almost limitless.'[3]

PMK-glycidate was seized in an Ecstasy and methamphetamine lab in the Netherlands in 2011 along with instructions for its conversion into PMK. PMK-glycidate remains legal in late 2012. While it does, MDMA supplies will remain abundant and of high quality. If international precursor conventions are again hastily amended, it's a fair bet that supplies will dwindle and replacements will be found. Whether that will be another analogue of a precursor or a new mephedrone-type drug remains to be seen.

Just as Chinese chemists have been evading the law for some years now by tweaking molecules to produce analogues of drugs, now the Russian, Italian and Israeli criminals who control the world's synthetic drugs markets are going one step further and producing legal analogues of banned precursor chemicals. It presents legislators with a fresh challenge: for each of the precursors that legislators ban, another will quickly appear. The circle closes like a ring substitution in Shulgin's laboratory, and this tale of unintended consequences closes with analogous, tail-chasing circularity.

# Notes

## Prologue: Contemporary Chemical Culture
1. www.urban75.net/forums/
   threads/6-apb-powder-some-mxe-blimey-this-interesting.294934/

## Chapter 1: Vegetable to Chemical
1. C. F. Gorman, 'Excavations at Spirit Cave, North Thailand: Some Interim Interpretations', *Asian Perspectives,* Vol. 13, 1970, pp. 79–108
2. www.antiquecannabisbook.com/chap2B/China/Pen-Tsao.htm
3. www.shipman-inquiry.org.uk/4r_page.asp?id=3107
4. Quoted in J. C. Poggendorff, *Annalen der Physik und Chemie* (Wiley VCH, 1828), Vol. 88, pp. 253–256; www.chem.yale.edu/~chem125/125/history99/4RadicalsTypes/UreaPaper1828.html
5. Frank J. Ayd, Jr. and Barry Blackwell, eds., *Discoveries in Biological Psychiatry* (J. B. Lippincott Company, 1970); www.psychedelic-library.org/hofmann.htm
6. William S. Burroughs and Allen Ginsberg, *The Yage Letters: Redux* (Penguin Modern Classics, 2008), p. 24
7. R. Gordon Wasson, 'Seeking The Magic Mushroom', *Life,* 13 May 1957
8. Havelock Ellis, 'Mescal, a New Artificial Paradise', *The Contemporary Review,* January 1898
9. www.independent.co.uk/arts-entertainment/music/news/revealed-dentist-who-introduced-beatles-to-lsd-415230.html
10. http://hansard.millbanksystems.com/lords/1966/aug/04/drugs-prevention-of-misuse-act-1964
11. www.presidency.ucsb.edu/ws/?pid=3047
12. http://hansard.millbanksystems.com/lords/1977/jun/20/misuse-of-drugs-act-1971-modification#S5LV038 4P0_19770620_HOL_148

## Chapter 2: The Great Ecstasy of the Toolmaker Shulgin

1. Werner Herzog's 1974 documentary *Die große Ekstase des Bildschnitzers Steiner* (The Great Ecstasy of the Sculptor Steiner) details the life and times of a champion ski-jumper who has so perfected his art he continually out-jumps the landing ramp; www.imdb.com/title/tt0070136/

2. Roland W. Freudenmann, Florian Öxler and Sabine Bernschneider-Reif, 'The Origin of MDMA (Ecstasy) Revisited: The True Story Reconstructed from the Original Documents', *Addiction*, Vol. 101, Issue 9, pp. 1241–1245; http://onlinelibrary.wiley.com/doi/10.1111/j.1360-0443.2006.01511.x/abstract

3. www.nytimes.com/2005/01/30/magazine/30ECSTASY.html?_r=1

4. Julian Palacios, *Syd Barrett and Pink Floyd: Dark Globe* (Plexus Publishing Ltd, 2010), p. 298

5. Dennis Romero, 'Sasha Shulgin, Psychedelic Chemist', *Los Angeles Times*, 5 September 1995

6. Alexander Shulgin, *PIHKAL: A Chemical Love Story* (Transform Press, 1991), p. 860

7. Ibid., p. xvi

8. Ibid., p. xviii

9. www.erowid.org/library/books_online/tihkal/shulgin_rating_scale.shtml

10. Shulgin, *PIHKAL*, pp. 876–877

11. Ibid., p. 733; see also www.erowid.org/library/books_online/PIHKAL109.shtml

12. www.maps.org/media/kleiman040204.html

13. Hugh Milne, 'Bhagwan, the God that Failed', cited in Matthew Collin, *Altered State* (Serpent's Tail, 1998), p. 33

14. Anthony D'Andrea, 'Ibiza: The Real Story of a Global Utopia', *Cultura*, Ibiza, Summer 2001, pp. 46–47; www3.ul.ie/sociology/index.php?pagid=23&memid=18

15. Peter Nasmyth, 'The Agony and the Ecstasy', *The Face*, October 1986, Issue 78, pp. 52–55

16. Nicholas Saunders, *E for Ecstasy* (Octavo, May 1993); http://ecstasy.org/books/e4x/e4x.ch.02.html

17. *The Face*, August 1990; http://testpressing.org/2010/07/the-face-europe-a-ravers-guide-august-1990/

18. www.ons.gov.uk/ons/rel/subnational-health3/deaths-related-to-drug-poisoning/2010/stb-deaths-related-to-drug-poisoning-2010.html

19. Simon Reynolds, *Rip it Up and Start Again: Postpunk 1978–1984* (Faber and Faber, 2006), p. xvi

## Chapter 3: The Birth of an Online Drugs Culture

1. John Markoff, *What the Dormouse Said: How the Sixties Counterculture Shaped*

*the Personal Computer Industry* (Penguin, 2005), p. 109

2.  Mylon Stolaroff, *Thanatos to Eros: 35 Years of Psychedelic Exploration: Ethnomedicine and the Study of Consciousness* (Thaneros Pr, 1994)

3.  http://kk.org/ct2/2008/09/the-whole-earth-blogalog.php

4.  Markoff, *What the Dormouse Said*, p. 109

5.  www.giganews.com/usenet-history/index.html

6.  Philip Elmer-Dewitt, 'First Nation in Cyberspace', *Time*, December 1993; www.time.com/time/magazine/article/0,9171,979768,00.html

7.  www.net.berkeley.edu/dcns/usenet/alt-creation-guide.html

8.  Steve Preisler, *Secrets of Methamphetamine Manufacture* (Loompanics, 1994)

9.  www.erowid.org/archive/rhodium/chemistry/eleusis/eleusis.vs.fester.html#Eleusis1

10. www.erowid.org/archive/rhodium/chemistry/eleusis/memoirs.html

11. www.erowid.org/psychoactives/faqs/faq_clandestine_chemistry.shtml

12. www.erowid.org/library/periodicals/journals/journals_telr.shtml

## Chapter 4: The Rise and Fall of the Research Chemical Scene

1.  Shulgin, *PIHKAL*, p. x

2.  Xuemei Huang, Danuta Marona-Lewicka and David E. Nichols, 'p-Methylthioamphetamine is a Potent New Non-Neurotoxic Serotonin-Releasing Agent', *European Journal of Pharmacology*, Vol. 229, Issue 1, December 1992, pp. 31–38; www.sciencedirect.com/science/article/pii/0014299992902829

3.  www.erowid.org/chemicals/4mta/4mta_info1.shtml

4.  jimmywoo, 'Flatlined Beyond Comprehension: experience with 4-MTA (ID 83154)', 30 April 2010, erowid.org/exp/83154

5.  www.erowid.org/psychoactives/law/countries/uk/uk_misuse_phen_2.shtml

6.  www.erowid.org/psychoactives/law/cases/federal/federal_analog1.shtml

7.  www.erowid.org/library/books_online/pihkal/pihkal043.shtml

8.  www.vice.com/read/criminal-chlorination-0000350-v19n9

9.  http://everything2.com/title/JLF+Poisonous+Non-Consumables

10. www.erowid.org/chemicals/2ct7/2ct7.shtml#deaths

11. 'Operation Pipe Dreams Puts 55 Illegal Drug Paraphernalia Sellers out of Business'; www.justice.gov/opa/pr/2003/February/03_crm_106.htm

12. www.justice.gov/dea/pubs/pressrel/pr072204.html

13. www.erowid.org/psychoactives/research_chems/research_chems_info1.shtml#dea_announcement

14. www.erowid.org/library/books_online/tihkal/tihkal37.shtml

15. news.bbc.co.uk/1/hi/england/4111625.stm

## Chapter 5: The Calm Before the Storm, and a Curious Drought

1. www.erowid.org/library/books_online/PIHKAL033.shtml
2. http://oreilly.com/web2/archive/what-is-web-20.html. The concept of 'Web 2.0' began with a conference brainstorming session between O'Reilly and MediaLive International. Dale Dougherty, web pioneer and O'Reilly VP, noted that far from having 'crashed', the web was more important than ever, with exciting new applications and sites popping up with surprising regularity. See also www.web2summit.com/web2011/public/content/about
3. www.imrg.org/ImrgWebsite/User/Pages/B2C_Global_e-Commerce_Overview_2011.aspx
4. Charlotte Walsh, 'Magic Mushrooms: From sacred entheogen to class A drug', *Entertainment and Sports Law Journal*, Vol. 4, No. 1, April 2006; go.warwick.ac.uk/eslj/issues/volume4/number1/walsh/. Walsh writes, 'The situation was further complicated by the conflicting interpretations of the law that emanated from government. Many of those who sold magic mushrooms used to display in their windows a photocopy of a letter, written by Home Office official Ian Breadmore in 2003, that clearly stated: "It is not illegal to sell or give away a freshly picked mushroom provided that it has not been prepared in any way."'
5. www.telegraph.co.uk/news/worldnews/europe/netherlands/3441105/Magic-mushrooms-banned-in-Netherlands.html
6. www.bbc.co.uk/news/uk-11816071
7. www.publications.parliament.uk/pa/cm200910/cmselect/cmhaff/74/7408.htm

## Chapter 6: Mephedrone Madness: the Underground Hits the High Street

1. Fauna & Flora International, press release, 2009: 'Destruction of "Ecstasy Oil Factories" deals severe blow to criminals in Phnom Samkos Wildlife Sanctuary'; www.flora-fauna.org. See also Adam Yamaguchi's video 'Forest of Ecstasy'; http://current.com/shows/vanguard/91315580_forest-of-ecstasy.htm
2. www.unodc.org/documents/data-and-analysis/WDR2011/World_Drug_Report_2011_ebook.pdf; data collated from pp. 33–40
3. Posting to the Hive bulletin board, found in archives of the site that now no longer exists. No live URL available
4. 'Gefährlicher Kick mit "Spice"', *Frankfurter Rundschau*, 12 December 2008; www.fr-online.de/rhein-main/frankfurt-gefaehrlicher-kick-mit--spice-,1472796,3375090.html
5. www.bluelight.ru/vb/threads/272660-Spice-Gold-Unbelievable/page5

6.  http://ericcarlin.wordpress.com/2010/04/02/
    my-acmd-resignation-letter-to-the-home-secretary/
7.  'A Collapse in Integrity of Scientific Advice in the UK', *The Lancet*, Vol. 375,
    Issue 9723, 17 April 2010, p. 1319; www.thelancet.com/journals/lancet/
    article/PIIS0140-6736(10)60556-9/fulltext

## Chapter 7: Woof Woof Is the New Meow Meow

1.  www.telegraph.co.uk/health/healthnews/7945058/Ivory-Wave-is-new-
    Miaow-Miaow.html
2.  '2011 Annual Report on the State of the Drugs Problem
    in Europe: New Drugs and Emerging Trends';
    www.emcdda.europa.eu/online/annual-report/2011/
    new-drugs-and-trends/2
3.  Simon D. Brandt, 'What Should Be Done about Mephedrone?', *British
    Medical Journal*, 16 June 2010; www.bmj.com/rapid-response/2011/11/02/
    analyses-second-generation-legal-highs-uk-confusing-case-nrg-1
4.  www.cambridge-news.co.uk/News/New-legal-high-to-hit-city-streets.
    htm
5.  www.dailymail.co.uk/home/moslive/article-1267582/The-Chinese-
    laboratories-scientists-work-new-meow-meow.html
6.  www.homeoffice.gov.uk/drugs/drug-law/temporary-class-drug-orders/
7.  The 1977 amendment to the 1971 Misuse of Drugs Act outlawed 'any
    compound (not being a compound for the time being specified in sub-
    paragraph (a) above) structurally derived from tryptamine or from a ring-
    hydroxy tryptamine by substitution at the nitrogen atom of the sidechain
    with one or more alkyl substituents but no other substituent.' AMT did not
    fit that precise definition, so was legal.
8.  www.bbc.co.uk/news/uk-scotland-tayside-central-14064996
9.  www.homeoffice.gov.uk/about-us/corporate-publications-strategy/
    home-office-circulars/circulars-2012/014-2012/
10. David E. Nichols and Stewart Frescas, 'Improvements to the Synthesis
    of Psilocybin and a Facile Method for Preparing the O-Acetyl Prodrug
    of Psilocin', *Synthesis*, No. 6, pp. 935–938 (1999); www.erowid.org/
    references/texts/show/6535docid6064
11. www.homeoffice.gov.uk/publications/science-research-statistics/
    research-statistics/crime-research/hosb1011/hosb1011?view=Binary
12. www.mixmag.net/drugssurvey
13. www.walesonline.co.uk/news/wales-news/2010/03/27/company-will-
    not-stop-handling-meow-meow-drug-payments-91466-26119009/
14. www.mixmag.net/drugssurvey

## Chapter 8: Ready-Salted Zombies and a Chemical Panic

1. Joan Miro, 'Thankful that I'm alive: experience with Bromo-dragonFLY (sold as 2C-B-Fly) (ID 81677)', 12 October 2009, erowid.org/exp/81677
2. Matthew A. Parker, Danuta Marona-Lewicka, Virginia L. Lucaites, David L. Nelson and David E. Nichols, 'A Novel (Benzodifuranyl)aminoalkane with Extremely Potent Activity at the 5-HT2A Receptor', *Journal of Medicinal Chemistry*, Vol. 41, 1998, pp. 5148–5149; www.unc.edu/~dlinz/Papers/A%20Novel%20(Benzodifuranyl)aminoalkane.pdf
3. http://media.carnegieinst.se/2012/04/NilsBejerot_inlaga_OK.pdf
4. http://ewsd.wiv-isp.be/Main/5-IT%20deaths%20in%20Sweden.aspx
5. www.bluelight.ru/vb/threads/274696-Describe-your-worst-psychedelic-experience(s)!?p=8658607#post8658607
6. www.flashback.org/sp27563892
7. www.kentonline.co.uk/kentonline/home/2012/march/15/hugo_wenn.aspx
8. www.bbc.co.uk/news/uk-england-leicestershire-17007766
9. http://metro.co.uk/2012/03/14/body-found-in-chingford-canal-believed-to-be-missing-partygoer-andrew-cooke-29-357437/
10. www.dailymail.co.uk/news/article-2121565/Will-learn-Mrs-Speaker-sparks-ANOTHER-storm-claiming-shes-tempted-try-Mexxy-drug-outlawed.html
11. www.startribune.com/local/north/135800088.html
12. www.huffingtonpost.com/2012/06/01/cdc-denies-zombies-existence_n_1562141.html?ref=tw
13. blogs.reuters.com/jackshafer/2012/05/31/drug-panics-bath-salts-and-face-eating-zombies/
14. www.huffingtonpost.com/subhash-kateel/its-bigger-than-bath-salts_b_1562014.html
15. Edward Huntington Williams MD, 'Negro cocaine fiends are a new southern menace', *New York Times*, 8 February 1914
16. Russell Russo, Noah Marks, Katy Morris, Heather King, Angelle Gelvin and Ronald Rooney, 'Life-threatening Necrotizing Fasciitis Due to "Bath Salts" Injection', *Orthopedics*, Vol. 35, Issue 1, January 2012
17. http://abcnewsradioonline.com/health-news/tag/bath-salts
18. www.nytimes.com/2011/10/11/us/states-adding-drug-test-as-hurdle-for-welfare.html?pagewanted=all
19. www.youtube.com/watch?feature=player_embedded&v=QgX4ICh1wQc

## Chapter 9: Your Crack's in the Post

1. Nicolas Christin, 'Traveling the Silk Road: A Measurement Analysis of a Large Anonymous Online Marketplace', Carnegie Mellon INI/CyLab, July 2012; http://arxiv.org/abs/1207.7139

2. www.nytimes.com/2005/06/12/opinion/12rich.html?pagewanted=all
3. Royal Mail Spokesman response quoted on www.bbc.co.uk/blogs/ watchdog/2012/05/royal_mail_1.html
4. Hidden in the code for the Genesis Block was this sentence, citing a *Times* of London report: 'The Times 03/Jan/2009 Chancellor on brink of second bailout for banks'.
5. www.wired.com/magazine/2011/11/mf_bitcoin/all/1
6. Ibid.
7. http://anonymity-in-bitcoin.blogspot.co.uk/2011/07/bitcoin-is-not-anonymous.html
8. www.philzimmermann.com/EN/essays/WhyIWrotePGP.html
9. www.theregister.co.uk/2009/11/24/ripa_jfl/page4.html
10. US Attorney's Office press release, 16 April 2012; www.justice.gov/usao/ cac/Pressroom/2012/045.html

## Chapter 10: Prohibition in the Digital Age

1. www.slate.com/articles/health_and_science/medical_examiner/2010/02/ the_chemists_war.html
2. Ibid.
3. www.prnewswire.com/news-releases/tell-your-children-they-can-get-naturally-high-177905211.html
4. www.healthland.time.com/2012/07/25/ after-the-ban-on-bath-salts-many-legal-highs-remain/#ixzz244z27e4z
5. www.nola.com/crime/index.ssf/2012/11/21-year-old_dies_after_one_dro. html
6. www.erowid.org/chemicals/2ci_nbome/2ci_nbome_death.shtml
7. 'New Drugs Detected in the EU at the Rate of Around One Per Week, Say Agencies'; www.emcdda.europa.eu/news/2012/2
8. Ibid.
9. www.kentonline.co.uk/kentonline/news/2012/february/9/cat_and_ mouse.aspx
10. www.ukba.homeoffice.gov.uk/sitecontent/newsarticles/2012/ february/36-drugs-guilty
11. www.soca.gov.uk/about-soca/library/doc_download/317-sars-annual-report-2011.pdf
12. www.homeoffice.gov.uk/publications/alcohol-drugs/drugs/ annual-review-drug-strategy-2010/acmd-letter-to-prof-les-iversen
13. Shulgin had a stroke in 2010. His stepdaughter, Wendy, emailed as this book was going to press with news that her stepfather's health had much improved. 'Sasha is doing well. He's had a wonderful sort of comeback from the severity of the dementia, due in part we believe to a medication that he's taking now which is ergotamine, or hydergine [a precursor for

LSD]. That has really helped him. He doesn't sundown much anymore, he's clearer and catches all conversation around him, following it, commenting on it, cracking jokes (we know he's well when he does that!). Between a good change in diet, more exercise, and that medication, he's really doing well.' A year before, I had asked Shulgin's wife Ann if her husband's illness might have been caused by his lifelong use of drugs. She was adamant it had not. 'Considering the hundreds of thousands of people who have experimented with psychoactive drugs and visionary plants, many of them using them as spiritual tools, there is no medical evidence whatsoever that that would be the case. It's simply not true,' she said.

14. www.homeoffice.gov.uk/publications/alcohol-drugs/drugs/
   annual-review-drug-strategy-2010/acmd-letter-to-prof-les-iversen
15. www.guardian.co.uk/commentisfree/2012/apr/07/
   latin-america-drugs-nightmare
16. www.guardian.co.uk/world/2011/nov/13/
   colombia-juan-santos-war-on-drugs
17. www.cdc.gov/nchs/fastats/smoking.htm
18. Sheryl Garrett, *Adventures in Wonderland: Decade of Club Culture* (Headline, 1999)
19. Matthew Collin, *Altered State* (Serpent's Tail, 1998)
20. 'New drugs detected in the EU at the rate of around one per week, say agencies'; www.emcdda.europa.eu/news/2012/2; see also 'Drug Policy Profiles: Portugal', Publications Office of the European Union (2011); and www.emcdda.europa.eu/attachements.cfm/att_137215_EN_PolicyProfile_ Portugal_WEB_Final.pdf, p. 24
21. 'Portugal drug law show results ten years on, experts say', AFP, 1 July 2011; www.google.com/hostednews/afp/article/ ALeqM5g9C6x99EnFVdFuXw_B8pvDRzLqcA
22. Independent Scientific Committee on Drugs, 2012: 'Analogue controls: An imperfect law'; www.ukdpc.org.uk/publication/ analogue-controls-an-imperfect-law/
23. David Fisher, '"Revolutionary" legal high law means state regulated drug market', www.nzherald.co.nz/nz/news/article. cfm?c_id=1&objectid=10822749

## Chapter 11: The End of the Road?

1. http://gawker.com/what-was-on-alleged-silk-road-boss-laptop-at-the-momen-1469122014/@adrianchen
2. https://docs.google.com/file/d/0Bzt0K7_O4qyqNjEtUG9USG5uQ3c/ edit?pli=1
3. http://www.scribd.com/doc/172773407/ Ulbricht-Criminal-Complaint-Silk-Road

4. http://www.scribd.com/doc/172773407/
Ulbricht-Criminal-Complaint-Silk-Road
5. http://www.shroomery.org/forums/showflat.php/Number/13860995
6. https://bitcointalk.org/index.php?topic=175.140%3Bwap2
7. http://gawker.com/
the-underground-website-where-you-can-buy-any-drug-imag-30818160
8. https://bitcointalk.org/index.php?topic=47811.msg568744#msg568744
9. http://www.linkedin.com/in/rossulbricht
10. http://www.amazon.com/Europium-Oxide-Thin-Films-Exploration-
Epitaxy/dp/3639174763/ref=sr_1_1?ie=UTF8&qid=1385740077&
sr=8-1&keywords=ross+ulbricht
11. http://www.liveleak.com/view?i=2b3_1380750540
12. https://www.rossulbricht.org/wp-content/uploads/2013/10/20131001-
Ross-Ulbricht-Silk-Road-Maryland-Grand-Jury-Indictment.pdf
13. http://grugq.tumblr.com/post/62986125689/
maryland-indictment-timeline; https://docs.google.com/
file/d/0BztOK7_O4qyqNjEtUG9USG5uQ3c/edit?pli=1
14. https://www.rossulbricht.org/wp-content/uploads/2013/10/20131001-
Ross-Ulbricht-Silk-Road-Maryland-Grand-Jury-Indictment.pdf
15. https://docs.google.com/file/d/0BztOK7_O4qyqNjEtUG9USG5uQ3c/
edit?pli=1
16. https://medium.com/matter/157af08832e6
17 http://www.reuters.com/article/2013/12/20/
us-silkroad-indictment-idUSBRE9BJ14H20131220
18. http://www.justice.gov/usao/nys/pressreleases/December13/
JonesetalArrestsSilkRoad2PR.php
19 http://www.reddit.com/r/DarkNetMarkets/
20 https://crypto.cat
21. Re: Coindesk.com Special Report: 56% of Bitcoiners Think Bitcoin Will
Reach US$10,000 in 2014.
22. http://www.coindesk.com/one-bitcoin-worth-ounce-gold-today/
23. http://www.reuters.com/article/2013/08/05/
us-dea-sod-idUSBRE97409R20130805
24. http://www.gwern.net/Silk%20Road
25. http://pastebin.com/CETqkjfu. Untitled pastebin document by a writer
claiming familiarity with every major drug-dealing site of the last decade.
26. http://www.theguardian.com/commentisfree/2014/jan/05/
left-silent-state-power-government-market
27. http://www.bbc.co.uk/news/world-us-canada-25598791
28. http://www.v3.co.uk/v3-uk/news/2280234/
prism-eu-approves-inquiry-into-nsa-cyber-snooping
29. http://www.wired.co.uk/news/archive/2013-07/27/pornwall

## Epilogue

1. euchinawto.org/index.php?option=com_ content&task=view &id=347 &Itemid=33&lang=eeu

2. 'Veel xtc-pillen bevatten hoge dosis', Associated Press, 31 January 2011; Document PZEEUC0020110131e71v0005n

3. www.unodc.org/unodc/en/frontpage/2012/March/limitless-ways-to-disguise-ways-to-manufacture-party-pills-unodc-expert.html

# Useful Organizations

For more information on drug policy, research, law, and issues around addiction, consult the following organizations:

**Release**

Providing free and confidential specialist advice to the public and professionals
www.release.org.uk

**Angelus Foundation**

Research, education and advocacy
www.angelusfoundation.com

***MixMag* Survey and Global Drugs Survey**

World's most in-depth drugs survey with focus on use among young people
http://globaldrugsurvey.com/run-my-survey/case-studies/mixmag

**Addaction**

Addresses dependent use of all drugs
www.addaction.org.uk

**Drugscope**

For independent information on drugs and drug-related issues
www.drugscope.org.uk

**Action on Addiction**

www.actiononaddiction.org.uk/home.aspx

**Transform**

Campaigning group working towards drug policy change
www.tdpf.org.uk

**The Club Drug Clinic**

For users of club drugs such as GHB, Ecstasy, mephedrone and
ketamine, and legal highs
www.clubdrugclinic.com

**Urban75**

Up-to-date information on drugs, effects and law. Impartial and
well-informed
www.urban75.com/Drugs/

**Pillreports**

Crowd-sourced Ecstasy quality reports
www.pillreports.com

**Bluelight**

Drugs forum with in-depth information around all new drugs
www.bluelight.ru

**Erowid**

Encyclopedic online drug library; essential reading
www.erowid.org

# Glossary

**Technical Terms**

Functional groups: Groups of atoms found within molecules that are involved in the chemical reactions characteristic of those molecules.

Ring substitution: To replace, using chemical reactions, atoms on the molecular structure of a chemical. In drug chemistry, this is deliberate and is carried out to change the drug's effects, or its legality.

Analogue: A contentious and legally complicated area; in some contexts this is taken to mean a chemical that is related structurally or in activity terms to a banned drug.

Reagent: Substance used to provoke a chemical reaction; in this book's context, to prove or disprove the presence of an illegal drug. An example is the Marquis reagent, used to demonstrate the presence of MDMA in pills sold as Ecstasy.

Precursor: A substance from which another is formed; in drug chemistry an example might be safrole or methylamine, used in MDMA manufacture.

Receptor binding: A measure of the efficacy of a drug in vivo; that is to say, how well a drug attaches to our brain and body's neurotransmitters.

**Types of Drugs**

Phenethylamines: Psychedelic and stimulant drugs, many of which were invented by Shulgin and published in *PIHKAL*. Most are somewhat similar in effect to mescaline, the natural psychedelic found in cacti. Most often in the context of designer drug use, these are the 2C-series of drugs, such as 2C-B and 2C-I, etc. Some, such as DOM and DOC, are super-potent, at a milligram or less.

Tryptamines: Psychedelic drugs most similar in effect to psilocybin-containing mushrooms.

Cathinones: Stimulants whose effects resemble those of amphetamines; the parent chemical is found in *Catha edulis*, or khat, the woody shrub found throughout eastern Africa, especially Yemen and Ethiopia. Mephedrone is a drug in this class, as are most US bath salts type drugs.

Piperidines: Super-potent stimulant class of drugs, to which Ritalin, the ADHD medicine, belongs. Long-lasting and low-dose.

Methamphetamines: Stimulant class of drugs related structurally and in effects profile to amphetamines, though often with a longer-lasting effect. Crystal meth, or 'ice', is the most common of these, and is popular among poorer drug users in the US.

Benzodiazepines: Tranquillizers, such as Valium (diazepam). There are hundreds of benzodiazepine variants possible and available.

NBOME-series: Super-potent psychedelic stimulants. Most often these structural variants on Shulgin's 2C-series of substituted phenethylamines are more potent by a factor of at least fifty. For example, 25I-NBOME is active at 300 μg, whereas 2C-I is active at 10 mg.

# Acknowledgements

With many thanks to . . .

Earth and Fire at Erowid, for your tireless work and pragmatism; to all at Pillreports for keeping people safe; to F&B (you know who you are); all posters at Bluelight and Urban75, especially editor Mike Slocombe for his contributions to harm reduction; to Yetman, Where Wolf, and Clapham Boy. To Agnetha Fältskog, wherever and whoever you may be.

To Max 'Narcomania' Daly at Drugscope for kicking it all off, John Ramsey at St George's for the analysis and commentary and unstinting generosity, Professor David E. Nichols for your generous interviews, and the Shulgin family for supporting the project and your permission to quote extensively from *PIHKAL*.

To Fiona Measham, Adam Winstock, Duncan Dick of *MixMag*, Danny Kushlick of Transform and Alex Stevens for generous interviews and commentary.

To Andrew Davies and Andrew Preston at *Live!* magazine for the commission that stepped things up a gear and to Tina Jackson for a very timely conversation. To Matthew Collin, for *Altered State* and many enlightening conversations.

Endless thanks to Andrew Gordon at David Higham Associates for the constant support, motivation, tact and diplomacy, and to Philip Gwyn Jones, Laura Barber and Michael Salu at Granta/Portobello for your enduring commitment to the project.

Special thanks go to Sara Holloway for a forensic and clear-sighted edit, and to all Portobello production staff, especially Jennie Condell for a meticulous copy-edit, and Christine Lo for making the virtual real.

To Frank Broughton and Jason Underhill for your wit, warmth and wisdom. Michael Cook for your words of encouragement that night when the story took a strange and important twist. To Bill Brewster for the parties and the music, and to Paul Byrne at testpressing.org for assistance and diversion.

To Jennifer Dunn for the chemistry lessons.

To Kiley & Nunn Builders, Canning Town's finest. Google them.

And to . . .

all those I have quoted who must remain anonymous

To John and Margaret Power, and our Phil, Chris, Jon and Andy.

Finally, and most fully, the biggest thank you of them all to Sasha Dunn, for your tolerance and logic and love.

mpx

# Index